# GLENCOE MATH

## BUILT TO THE COMMON CORE (CCSS)

**AUTHORS**
Carter • Cuevas • Day • Malloy
Kersaint • Reynosa • Silbey • Vielhaber

Mc
Graw
Hill
Education

Bothell, WA • Chicago, IL • Columbus, OH • New York, NY

**connectED.mcgraw-hill.com**

Send all inquiries to:
McGraw-Hill Education
8787 Orion Place
Columbus, OH  43240

ISBN:  978-0-02-145425-9 (*Volume 2*)
MHID:  0-02-145425-6

Printed in the United States of America.

6 7 8 9 QSX 20 19 18 17 16

# CONTENTS IN BRIEF

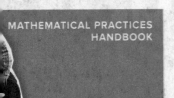

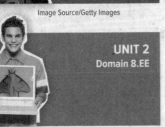

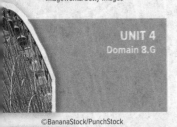

# Everything you need,

# anytime, anywhere.

With ConnectED, you have instant access to all of your study materials—anytime, anywhere. From homework materials to study guides—it's all in one place and just a click away. ConnectED even allows you to collaborate with your classmates and use mobile apps to make studying easy.

## Resources built for you—available 24/7:

- Your eBook available wherever you are

- Personal Tutors and Self-Check Quizzes to help your learning

- An Online Calendar with all of your due dates

- eFlashcard App to make studying easy

- A message center to stay in touch

## Go Mobile!

Visit mheonline.com/apps to get entertainment, instruction, and education on the go with ConnectED Mobile and our other apps available for your device.

## Go Online!

connectED.mcgraw-hill.com

your Username

your Password

**Vocab**

Learn about new vocabulary words.

**Watch**

Watch animations and videos.

**Tutor**

See and hear a teacher explain how to solve problems.

**Tools**

Explore concepts with virtual manipulatives.

**Sketchpad**

Discover concepts using The Geometer's Sketchpad®.

**Check**

Check your progress.

**eHelp**

Get targeted homework help.

**Worksheets**

Access practice worksheets.

# Chapter 1
# Real Numbers

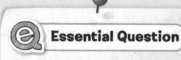

**Essential Question**

WHY is it helpful to write numbers in different ways?

Real World
p. 89

# UNIT PROJECT 103

**Music to My Ears**

# Chapter 2
# Equations in One Variable

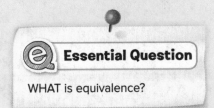

**Essential Question**

WHAT is equivalence?

Real World
p. 129

# Chapter 3
# Equations in Two Variables

**e** **Essential Question**

WHY are graphs helpful?

**Web Design 101**

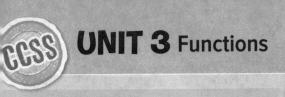

# UNIT 3 Functions

UNIT PROJECT PREVIEW
page 262

# Chapter 4
# Functions

 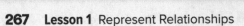

**Essential Question**

HOW can we model relationships between quantities?

Real World
p. 327

# UNIT PROJECT 361

### Green Thumb

# Chapter 5
# Triangles and the Pythagorean Theorem

**What Tools Do You Need?** 366
**What Do You Already Know?** 367
**Are You Ready?** 368

**Essential Question**

HOW can algebraic concepts be applied to geometry?

Real World
p. 431

# Chapter 6
# Transformations

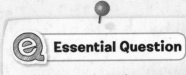

**Essential Question**

HOW can we best show or
describe the change in
position of a figure?

Real World
p. 487

# Chapter 7
# Congruence and Similarity

**Essential Question**

HOW can you determine congruence and similarity?

p. 545

# Chapter 8
# Volume and Surface Area

### Essential Question

WHY are formulas important in math and science?

Real World
p. 597

## UNIT PROJECT 655

### Design That Ride!

xiii

UNIT PROJECT PREVIEW
page 658

# Chapter 9
# Scatter Plots and Data Analysis

Real World
p. 701

◉ **Essential Question**

HOW are patterns used when comparing two quantities?

# UNIT PROJECT 731

## Olympic Games

# Common Core State Standards for MATHEMATICS, Grade 8

*Glencoe Math,* Course 3, focuses on three critical areas: (1) applying equations in one and two variables; (2) understanding the concept of a function and using functions to describe quantitative relationships; (3) applying the Pythagorean Theorem and the concepts of similarity and congruence.

## Content Standards

**Domain 8.NS**

### The Number System
- Know that there are numbers that are not rational, and approximate them by rational numbers.

**Domain 8.EE**

### Expressions and Equations
- Work with radicals and integer exponents.
- Understand the connections between proportional relationships, lines, and linear equations.
- Analyze and solve linear equations and pairs of simultaneous linear equations.

**Domain 8.F**

### Functions
- Define, evaluate, and compare functions.
- Use functions to model relationships between quantities.

**Domain 8.G**

### Geometry
- Understand congruence and similarity using physical models, transparencies, or geometry software.
- Understand and apply the Pythagorean Theorem.
- Solve real-world and mathematical problems involving volume of cylinders, cones and spheres.

**Domain 8.SP**

### Statistics and Probability
- Investigate patterns of association in bivariate data.

## ⓂⓅ Mathematical Practices

1   Make sense of problems and persevere in solving them.
2   Reason abstractly and quantitatively.
3   Construct viable arguments and critique the reasoning of others.
4   Model with mathematics.
5   Use appropriate tools strategically.
6   Attend to precision.
7   Look for and make use of structure.
8   Look for and express regularity in repeated reasoning.

# Track Your Common Core Progress

These pages list the key ideas that you should be able to understand by the end of the year. You will rate how much you know about each one. Don't worry if you have no clue **before** you learn about them. Watch how your knowledge grows as the year progresses!

 I have no clue.      I've heard of it.     I know it!

|  | Before | | | After | | |
|---|---|---|---|---|---|---|
| **8.NS   The Number System** | | | | | | |
| **Know that there are numbers that are not rational, and approximate them by rational numbers.** | | | | | | |
| 8.NS.1  Know that numbers that are not rational are called irrational. Understand informally that every number has a decimal expansion; for rational numbers show that the decimal expansion repeats eventually, and convert a decimal expansion which repeats eventually into a rational number. | | | | | | |
| 8.NS.2  Use rational approximations of irrational numbers to compare the size of irrational numbers, locate them approximately on a number line diagram, and estimate the value of expressions (e.g., $\pi^2$). | | | | | | |
| **8.EE   Expressions and Equations** | | | | | | |
| **Work with radicals and integer exponents.** | | | | | | |
| 8.EE.1  Know and apply the properties of integer exponents to generate equivalent numerical expressions. | | | | | | |
| 8.EE.2  Use square root and cube root symbols to represent solutions to equations of the form $x^2 = p$ and $x^3 = p$, where $p$ is a positive rational number. Evaluate square roots of small perfect squares and cube roots of small perfect cubes. Know that $\sqrt{2}$ is irrational. | | | | | | |
| 8.EE.3  Use numbers expressed in the form of a single digit times an integer power of 10 to estimate very large or very small quantities, and to express how many times as much one is than the other. | | | | | | |
| 8.EE.4  Perform operations with numbers expressed in scientific notation, including problems where both decimal and scientific notation are used. Use scientific notation and choose units of appropriate size for measurements of very large or very small quantities (e.g., use millimeters per year for seafloor spreading). Interpret scientific notation that has been generated by technology. | | | | | | |
| **Understand the connections between proportional relationships, lines, and linear equations.** | | | | | | |
| 8.EE.5  Graph proportional relationships, interpreting the unit rate as the slope of the graph. Compare two different proportional relationships represented in different ways. | | | | | | |
| 8.EE.6  Use similar triangles to explain why the slope $m$ is the same between any two distinct points on a non-vertical line in the coordinate plane; derive the equation $y = mx$ for a line through the origin and the equation $y = mx + b$ for a line intercepting the vertical axis at $b$. | | | | | | |

|  | Before | | | After | | |
|---|---|---|---|---|---|---|
| **8.EE   Expressions and Equations** *continued* | 😞 | 😐 | 🙂 | 😞 | 😐 | 🙂 |
| **Analyze and solve linear equations and pairs of simultaneous linear equations.** | | | | | | |
| **8.EE.7**  Solve linear equations in one variable.<br>  **a.** Give examples of linear equations in one variable with one solution, infinitely many solutions, or no solutions. Show which of these possibilities is the case by successively transforming the given equation into simpler forms, until an equivalent equation of the form $x = a$, $a = a$, or $a = b$ results (where $a$ and $b$ are different numbers).<br>  **b.** Solve linear equations with rational number coefficients, including equations whose solutions require expanding expressions using the distributive property and collecting like terms. | | | | | | |
| **8.EE.8**  Analyze and solve pairs of simultaneous linear equations.<br>  **a.** Understand that solutions to a system of two linear equations in two variables correspond to points of intersection of their graphs, because points of intersection satisfy both equations simultaneously.<br>  **b.** Solve systems of two linear equations in two variables algebraically, and estimate solutions by graphing the equations. Solve simple cases by inspection.<br>  **c.** Solve real-world and mathematical problems leading to two linear equations in two variables. | | | | | | |

|  | Before | | | After | | |
|---|---|---|---|---|---|---|
| **8.F   Functions** | 😞 | 😐 | 🙂 | 😞 | 😐 | 🙂 |
| **Define, evaluate, and compare functions.** | | | | | | |
| **8.F.1**  Understand that a function is a rule that assigns to each input exactly one output. The graph of a function is the set of ordered pairs consisting of an input and the corresponding output. | | | | | | |
| **8.F.2**  Compare properties of two functions each represented in a different way (algebraically, graphically, numerically in tables, or by verbal descriptions). | | | | | | |
| **8.F.3**  Interpret the equation $y = mx + b$ as defining a linear function, whose graph is a straight line; give examples of functions that are not linear. | | | | | | |
| **Use functions to model relationships between quantities.** | | | | | | |
| **8.F.4**  Construct a function to model a linear relationship between two quantities. Determine the rate of change and initial value of the function from a description of a relationship or from two $(x, y)$ values, including reading these from a table or from a graph. Interpret the rate of change and initial value of a linear function in terms of the situation it models, and in terms of its graph or a table of values. | | | | | | |
| **8.F.5**  Describe qualitatively the functional relationship between two quantities by analyzing a graph (e.g., where the function is increasing or decreasing, linear or nonlinear). Sketch a graph that exhibits the qualitative features of a function that has been described verbally. | | | | | | |

|  | Before | | | After | | |
|---|---|---|---|---|---|---|
| **8.G   Geometry** | 😞 | 😐 | 🙂 | 😞 | 😐 | 🙂 |
| **Understand congruence and similarity using physical models, transparencies, or geometry software.** | | | | | | |
| **8.G.1**  Verify experimentally the properties of rotations, reflections, and translations:<br>  **a.** Lines are taken to lines, and line segments to line segments of the same length.<br>  **b.** Angles are taken to angles of the same measure.<br>  **c.** Parallel lines are taken to parallel lines. | | | | | | |

| | Before | | | After | | |
|---|---|---|---|---|---|---|

## 8.G Geometry continued

**8.G.2** Understand that a two-dimensional figure is congruent to another if the second can be obtained from the first by a sequence of rotations, reflections, and translations; given two congruent figures, describe a sequence that exhibits the congruence between them.

**8.G.3** Describe the effect of dilations, translations, rotations, and reflections on two-dimensional figures using coordinates.

**8.G.4** Understand that a two-dimensional figure is similar to another if the second can be obtained from the first by a sequence of rotations, reflections, translations, and dilations; given two similar two-dimensional figures, describe a sequence that exhibits the similarity between them.

**8.G.5** Use informal arguments to establish facts about the angle sum and exterior angle of triangles, about the angles created when parallel lines are cut by a transversal, and the angle-angle criterion for similarity of triangles.

**Understand and apply the Pythagorean Theorem.**

**8.G.6** Explain a proof of the Pythagorean Theorem and its converse.

**8.G.7** Apply the Pythagorean Theorem to determine unknown side lengths in right triangles in real-world and mathematical problems in two and three dimensions.

**8.G.8** Apply the Pythagorean Theorem to find the distance between two points in a coordinate system.

**Solve real-world and mathematical problems involving volume of cylinders, cones, and spheres.**

**8.G.9** Know the formulas for the volumes of cones, cylinders, and spheres and use them to solve real-world and mathematical problems.

## 8.SP Statistics and Probability

**Investigate patterns of association in bivariate data.**

**8.SP.1** Construct and interpret scatter plots for bivariate measurement data to investigate patterns of association between two quantities. Describe patterns such as clustering, outliers, positive or negative association, linear association, and nonlinear association.

**8.SP.2** Know that straight lines are widely used to model relationships between two quantitative variables. For scatter plots that suggest a linear association, informally fit a straight line, and informally assess the model fit by judging the closeness of the data points to the line.

**8.SP.3** Use the equation of a linear model to solve problems in the context of bivariate measurement data, interpreting the slope and intercept.

**8.SP.4** Understand that patterns of association can also be seen in bivariate categorical data by displaying frequencies and relative frequencies in a two-way table. Construct and interpret a two-way table summarizing data on two categorical variables collected from the same subjects. Use relative frequencies calculated for rows or columns to describe possible association between the two variables.

# UNIT 4
# CCSS Geometry

## Essential Question
HOW can you use different measurements to solve real-life problems?

### Chapter 5
## Triangles and the Pythagorean Theorem

In this chapter, you will use the Pythagorean Theorem to find side lengths of right triangles and distances on the coordinate plane.

### Chapter 6
## Transformations

In this chapter, you will describe the effect of translations, reflections, rotations, and dilations on geometric figures.

### Chapter 7
## Congruence and Similarity

In this chapter, you will describe transformations that produce congruent and similar figures.

### Chapter 8
## Volume and Surface Area

In this chapter, you will find the volume and surface area of cones, cylinders, and spheres.

# Unit Project Preview

**Design That Ride** Do you remember how you felt the first time you were on a roller coaster? Scared, excited, terrified? Amusement park rides come in all shapes and sizes and are thrilling to ride.

Although amusement park rides are fun, they are also designed to be very safe. Engineers apply geometric concepts and use precise measurements when designing these rides.

At the end of Chapter 8, you'll complete a project to find how mathematical concepts are used to design an amusement park ride. But for now, it's time to do an activity in your book. Sketch an amusement park ride. Label some geometric shapes that can be found within your ride.

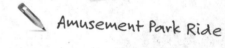

Amusement Park Ride

# Chapter 5
# Triangles and the Pythagorean Theorem

## Essential Question

HOW can algebraic concepts be applied to geometry?

## Common Core State Standards

**Content Standards**
8.G.5, 8.G.6, 8.G.7, 8.G.8, 8.EE.2

MP **Mathematical Practices**
1, 2, 3, 4, 5, 7, 8

## Math in the Real World

**Games** At a park in Morro Bay, California, one of the world's largest chess boards is made of concrete and has an area of 256 square feet. The chess pieces used weigh between 18 and 30 pounds each.

Label the dimensions of the chess board and one of the squares.

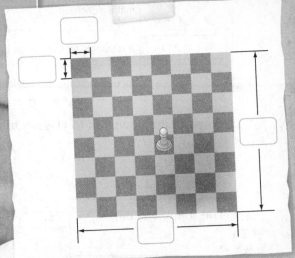

## FOLDABLES
### Study Organizer

 Cut out the Foldable on page FL3 of this book.

 Place your Foldable on page 442.

 Use the Foldable throughout this chapter to help you learn about the Pythagorean Theorem.

**Vocabulary**

| | | |
|---|---|---|
| alternate exterior angles | hypotenuse | proof |
| alternate interior angles | inductive reasoning | Pythagorean Theorem |
| converse | informal proof | regular polygon |
| corresponding angles | interior angles | remote interior angles |
| deductive reasoning | legs | theorem |
| Distance Formula | paragraph proof | transversal |
| equiangular | parallel lines | triangle |
| exterior angles | perpendicular lines | two-column proof |
| formal proof | polygon | |

## Study Skill: The Structure of Math

**Use a Flowchart** A *flowchart* is like a map that tells you how to get from the beginning of a problem to the end.

| Flowchart Symbols | |
|---|---|
| ◇ | A diamond contains a question. You need to stop and make a decision. |
| ▭ | A rectangle tells you what to do. |
| ⬭ | An oval indicates the beginning or end. |

Complete the flowchart to classify triangles by their sides.

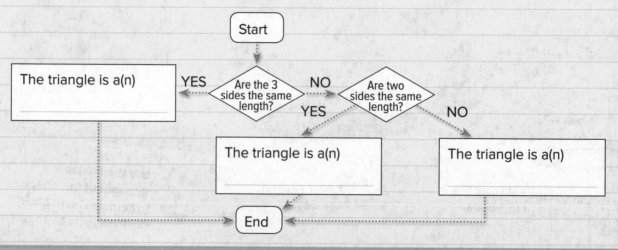

# What Do You Already Know?

List three things you already know about lines, angles, and triangles in the first section. Then list three things you would like to learn about lines, angles, and triangles in the second section.

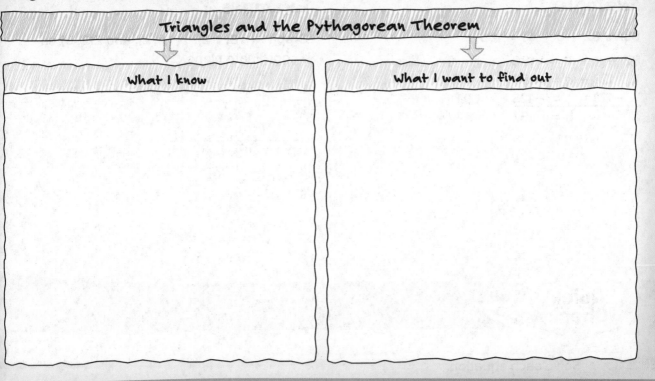

Triangles and the Pythagorean Theorem

| What I know | What I want to find out |
|---|---|
| | |

# When Will You Use This?

Here are a few examples of how angles and lines are used in the real world.

**Activity 1** Have you ever built a ramp for something? In the space below, draw a bicycle ramp that you would like to build. Be sure to include measurements for the lengths of the side of the ramp.

_____

_____

_____

Mandar and Roberto in
**Ramping Up**

Bike Ramp Plans

Side View

152°

50°

10 ft

Looks pretty simple.

Uh-oh...

**Activity 2** Go online at **connectED.mcgraw-hill.com** to read the graphic novel *Ramping Up*. What are the measures of the known angles in the plan for the bike ramp? _____

 **Are You Ready?**

Try the Quick Check below.
Or, take the Online Readiness Quiz.

 Check ✓

**CCSS** **Quick Review**

**Common Core Review** 6.EE.7, 6.NS.6

## Example 1

**Solve $82 + g + 41 = 180$**

$82 + g + 41 = 180$    Write the equation.

$123 + g = 180$    Add 82 and 41.

$\underline{-123 \qquad = -123}$    Subtraction Property of Equality

$g = 57$

## Example 2

**Graph $X(-3, -2)$, $Y(-1, 2)$, and $Z(2, 3)$ on a coordinate plane.**

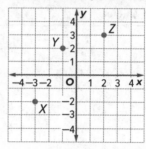

Start at the origin. The first number in each ordered pair is the x-coordinate. The second number in each ordered pair is the y-coordinate.

## Quick Check

**Equations** Solve each equation.

1. $49 + b + 45 = 180$

2. $t + 98 + 55 = 180$

3. $15 + 67 + k = 180$

 Show your work.

**Coordinate Plane** Graph and label each point on the coordinate plane.

4. $A(2, 4)$

5. $B(-1, -3)$

6. $C(0, -4)$

7. $D(3, -2)$

8. $E(-4, -3)$

9. $F(3, 0)$

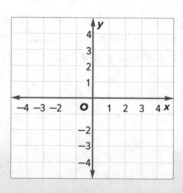

**How Did You Do?**

Which problems did you answer correctly in the Quick Check? Shade those exercise numbers below.

 1   2   3    4    5   6    7    8    9

# Inquiry Lab
## Parallel Lines

**Inquiry** WHAT are the angle relationships formed when a third line intersects two parallel lines?

 **Content Standards**
8.G.5

 **Mathematical Practices**
1, 3, 5

A newspaper route has two parallel streets. The streets are cut by another street as shown in the figure below.

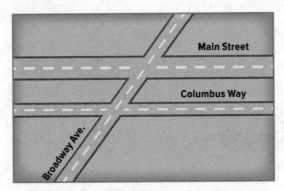

# Hands-On Activity

Parallel lines have special angle relationships. You will examine those relationships in this activity.

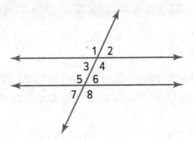

**Step 1** Use a protractor and angle relationships you have previously learned to find the measure of each numbered angle and record it in the table.

| Angle | 1 | 2 | 3 | 4 | 5 | 6 | 7 | 8 |
|---|---|---|---|---|---|---|---|---|
| Measure | | | | | | | | |

**Step 2** Color the angles that have the same measure.

**Step 3** Describe the position of the angles with the same measure.

_____

_____

# Investigate

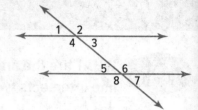

**MP** **Use Math Tools** Work with a partner. If the measure of ∠1 in the figure at the right is 40°, determine the measure of each given angle without using a protractor. Then check your answers by measuring with a protractor.

1. ∠2 _____

2. ∠3 _____

3. ∠4 _____

4. ∠5 _____

5. ∠6 _____

6. ∠7 _____

7. ∠8 _____

# Analyze and Reflect

**Refer to the figure above.**

8. What is the relationship between the two horizontal lines?

_____

_____

9. What is true about the measures of angles that are side by side?

_____

_____

10. **MP** **Reason Inductively** Congruent angles are angles that have the same measure. Describe the position of the congruent angles.

_____

_____

_____

# Create

11. **MP** **Make a Conjecture** Draw a set of parallel lines cut by another line. Estimate the measures of the eight angles formed. Check your estimates by measuring each angle with a protractor.

12. **Inquiry** WHAT are the angle relationships formed when a third line intersects two parallel lines?

_____

_____

_____

# Lines

## Vocabulary Start-Up

When two lines intersect in a plane and form right angles they are called **perpendicular lines**. Two lines are called **parallel lines** when they are in the same plane and do not intersect.

**Complete the graphic organizer.**

|  | Parallel Lines | Perpendicular Lines |
|---|---|---|
| Symbols | ‖ | ⊥ |
| Define it in your own words |  |  |
| Draw it |  |  |
| Describe a real-world example of it |  |  |

 **Essential Question**

HOW can algebraic concepts be applied to geometry?

**Vocabulary**

perpendicular lines
parallel lines
transversal
interior angles
exterior angles
alternate interior angles
alternate exterior angles
corresponding angles

**Math Symbols**
‖ is parallel to
⊥ is perpendicular to
$m\angle 1$ the measure of $\angle 1$

**CCSS** **Common Core State Standards**

**Content Standards**
8.G.5
**MP** **Mathematical Practices**
1, 3, 4

##  Real-World Link

A gymnastic event in the Summer Olympics involves the parallel bars. The women compete on uneven parallel bars and the men compete on the parallel bars like the one shown. Circle the parallel lines shown in the photo at the right.

**Which MP Mathematical Practices did you use?**
**Shade the circle(s) that applies.**

① Persevere with Problems
② Reason Abstractly
③ Construct an Argument
④ Model with Mathematics
⑤ Use Math Tools
⑥ Attend to Precision
⑦ Make Use of Structure
⑧ Use Repeated Reasoning

# Transversals and Angles

A line that intersects two or more lines is called a **transversal**, and eight angles are formed.

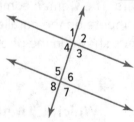

**Interior angles** lie inside the lines.
**Examples:** ∠3, ∠4, ∠5, ∠6

**Exterior angles** lie outside the lines.
**Examples:** ∠1, ∠2, ∠7, ∠8

**Alternate interior angles** are interior angles that lie on opposite sides of the transversal. When the lines are parallel, their measures are equal. **Examples:** $m\angle 4 = m\angle 6$, $m\angle 3 = m\angle 5$

**Alternate exterior angles** are exterior angles that lie on opposite sides of the transversal. When the lines are parallel, their measures are equal. **Examples:** $m\angle 1 = m\angle 7$, $m\angle 2 = m\angle 8$

**Corresponding angles** are those angles that are in the same position on the two lines in relation to the transversal. When the lines are parallel, their measures are equal. **Examples:** $m\angle 1 = m\angle 5$, $m\angle 2 = m\angle 6$, $m\angle 4 = m\angle 8$, $m\angle 3 = m\angle 7$

## Angles

Read $m\angle 1$ as the measure of angle 1.

## Parallel and Perpendicular Lines

Read $m \perp n$ as line $m$ is perpendicular to line $n$.

Read $p \parallel q$ as line $p$ is parallel to line $q$.

Special notation is used to indicate perpendicular and parallel lines.

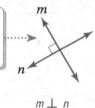

A red right angle symbol indicates that lines $m$ and $n$ are perpendicular.

$m \perp n$

Red arrowheads indicate that lines $p$ and $q$ are parallel.

$p \parallel q$

## Examples

Tutor

Classify each pair of angles in the figure as *alternate interior*, *alternate exterior*, or *corresponding*.

**1.** ∠1 and ∠7

∠1 and ∠7 are exterior angles that lie on opposite sides of the transversal. They are alternate exterior angles.

**2.** ∠2 and ∠6

∠2 and ∠6 are in the same position on the two lines. They are corresponding angles.

## Got It? Do this problem to find out.

a. Classify the relationship between ∠4 and ∠6. Explain.

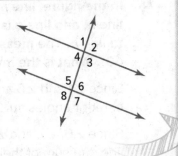

a. _____

# Find Missing Angle Measures

When two parallel lines are cut by a transversal, special angle relationships exist. If you know the measure of one of the angles, you can find the measures of all of the angles. Suppose you know that $m\angle 1 = 50°$. You can use that to find the measures of angles 2, 3, and 4.

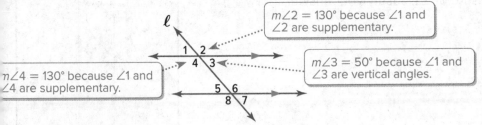

$m\angle 2 = 130°$ because ∠1 and ∠2 are supplementary.

$m\angle 3 = 50°$ because ∠1 and ∠3 are vertical angles.

$m\angle 4 = 130°$ because ∠1 and ∠4 are supplementary.

**STOP and Reflect**

In the figure, how do you know that $m\angle 5 = 50°$? Explain below.

## Example

**3.** A furniture designer built the bookcase shown. Line $a$ is parallel to line $b$. If $m\angle 2 = 105°$, find $m\angle 6$ and $m\angle 3$. Justify your answer.

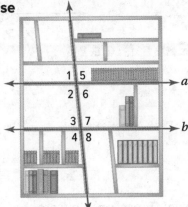

Since ∠2 and ∠6 are supplementary, the sum of their measures is 180°.
$m\angle 6 = 180° - 105°$ or 75°

Since ∠6 and ∠3 are interior angles that lie on opposite sides of the transversal, they are alternate interior angles. The measures of alternate interior angles are equal. $m\angle 3 = 75°$

## Got It? Do this problem to find out.

b. Refer to the situation above. Find $m\angle 4$. Justify your answer.

b. _____

## Example

**4.** In the figure, line *m* is parallel to line *n*, and line *q* is perpendicular to line *p*. The measure of ∠1 is 40°. What is the measure of ∠7?

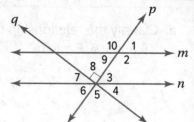

Since ∠1 and ∠6 are alternate exterior angles, m∠6 = 40°.

Since ∠6, ∠7, and ∠8 form a straight line, the sum of their measures is 180°.

40 + 90 + m∠7 = 180

So, m∠7 is 50°.

## Guided Practice

Check

**1.** Refer to the porch stairs shown. Line *m* is parallel to line *n* and m∠7 is 35°. Find the measure of ∠1. Justify your answer. (Example 3)

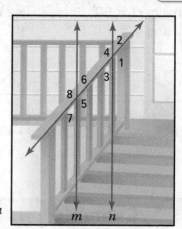

_____

_____

_____

**Refer to the figure at the right. Line *a* is parallel to line *b* and m∠2 is 135°. Find each given angle measure. Justify your answer.** (Examples 1, 2, and 4)

**2.** m∠9 _____

_____

**3.** m∠7 _____

_____

**4.**  **Building on the Essential Question** How are the measures of the angles related when parallel lines are cut by a transversal?

_____

_____

### Rate Yourself!

How confident are you about lines and angles? Check the box that applies.

For more help, go online to access a Personal Tutor.

Tutor

# Independent Practice

Go online for Step-by-Step Solutions

eHelp

**Classify each pair of angles as *alternate interior*, *alternate exterior*, or *corresponding*.** (Examples 1 and 2)

**1.** ∠2 and ∠4 _____

**2.** ∠4 and ∠5 _____

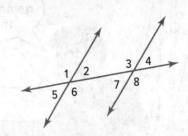

**3** In the flag shown at the right, line *a* is parallel to line *b*. If $m\angle 1 = 150°$, find $m\angle 4$ and $m\angle 7$. Justify

your answers. (Example 3) _____

_____

_____

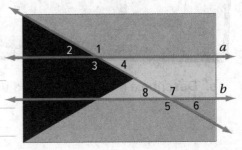

**Refer to the figure at the right. Line *s* is parallel to line *t*, $m\angle 2$ is 110° and $m\angle 11$ is 137°. Find each given angle measure. Justify your answer.** (Example 4)

**4.** $m\angle 7$ _____

_____

**5.** $m\angle 8$ _____

_____

**6.** $m\angle 3$ _____

_____

**7.** The parallel lines at the right are cut by a transversal. Find the value of *x*.

  **a.** Angles 1 and 2 are corresponding angles, $m\angle 1 = 45°$,

  and $m\angle 2 = (x + 25)°$. _____

  **b.** Angles 3 and 4 are alternate interior angles, $m\angle 3 = 2x°$,

  and $m\angle 4 = 80°$. _____

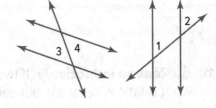

**8.** Describe a method you could use to find the value of *x* in the figure at the right without using a protractor.

_____

_____

_____

**9** **MP** **Model with Mathematics** Refer to the graphic novel frame below for Exercises a–b.

**a.** Describe a method you could use to find the missing angle.

_____

_____

**b.** Use your method from part **a** to find the measure of the missing angle.

_____

## H.O.T. Problems Higher Order Thinking

**10.** **MP** **Persevere with Problems** Quadrilateral *ABCD* is a parallelogram. Make a conjecture about the relationship of

∠1 and ∠2. Justify your reasoning. _____

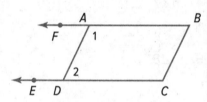

_____

_____

_____

_____

**11.** **MP** **Reason Inductively** If two parallel lines are cut by a transversal, what relationship exists between interior angles that are on the same side

of the transversal? _____

**12.** **MP** **Reason Inductively** Suppose *m*∠1 = *x*°. Use an informal argument to write an expression for the measure of ∠6 in the diagram at the right.

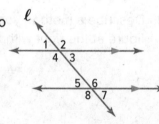

_____

_____

_____

# Extra Practice

**Classify each pair of angles as *alternate interior*, *alternate exterior*, or *corresponding*.**

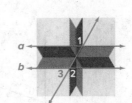

**13.** ∠3 and ∠6 _alternate interior_

∠3 and ∠6 are interior angles that lie on opposite sides of

the transversal. They are alternate interior angles

**14.** ∠1 and ∠3 _____

**15.** ∠2 and ∠7 _____

**16.** In the quilt design at the right, line *a* is parallel to line *b*.
If $m\angle1 = 120°$, find $m\angle2$ and $m\angle3$.

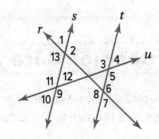

Justify your answers. _____

_____

_____

_____

**Refer to the figure at the right. Line *s* is parallel to line *t*, $m\angle2$ is 110° and $m\angle11$ is 137°. Find each given angle measure. Justify your answer.**

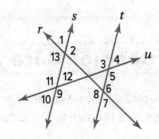

**17.** $m\angle6$ _____

_____

**18.** $m\angle13$ _____

_____

**19.** $m\angle4$ _____

_____

**20.** **MP** **Model with Mathematics** Draw a pair of parallel
lines cut by a transversal. Estimate the measure of one angle
and label it. Without using a protractor, label all the other angles
with their approximate measure.

**21.** Lines *m* and *n* are parallel and cut by the transversal *p*. Which of the following pairs of angles represent corresponding angles? Select all that apply.

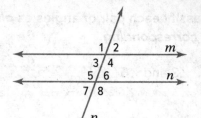

☐ ∠2 and ∠6

☐ ∠4 and ∠6

☐ ∠3 and ∠4

☐ ∠1 and ∠5

**22.** Lines *a* and *b* are parallel and cut by the transversal *t*. Label each of the 7 unknown angles with the correct angle measure.

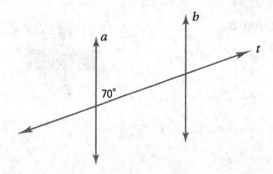

**23.** A poster has a triangular image with a base that measure 4 inches, and a height that measures 8 inches. What is the area of the poster? **6.G.1**

_____

Classify each pair of angles as *complementary*, *supplementary*, or *neither*. **7.G.5**

**24.** _____

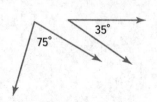

**25.** _____

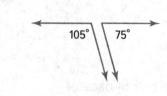

**26.** _____

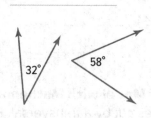

## Real-World Link

 Watch

**Detectives** A police detective uses analytical thinking to solve crimes. **Inductive reasoning** is the process of making a conjecture after observing several examples.

Unlike inductive reasoning, **deductive reasoning** uses facts, rules, definitions, or laws to make conjectures from given situations.

**Complete the graphic organizer by matching each situation with the type of reasoning used.**

Every time Bill watches his favorite team on television, the team loses. So, he decides to not watch the team play on TV.

Deductive Reasoning

In order to play sports, you need to have a B average. Simon has a B average, so he concludes he can play sports.

All triangles have 3 sides and 3 angles. Mariah has a figure with 3 sides and 3 angles, so it must be a triangle.

After performing a science experiment, LaDell concluded that only 80% of tomato seeds would grow into plants.

Inductive Reasoning

## Essential Question

HOW can algebraic concepts be applied to geometry?

### Vocabulary

inductive reasoning
deductive reasoning
proof
paragraph proof
informal proof
two-column proof
formal proof
theorem

### Common Core State Standards

**Content Standards**
Preparation for 8.G.6

**MP Mathematical Practices**
1, 2, 3, 4

### Which MP **Mathematical Practices** did you use?
### Shade the circle(s) that applies.

① Persevere with Problems
② Reason Abstractly
③ Construct an Argument
④ Model with Mathematics

⑤ Use Math Tools
⑥ Attend to Precision
⑦ Make Use of Structure
⑧ Use Repeated Reasoning

# The Proof Process

**Step 1** List the given information, or what you know. If possible, draw a diagram to illustrate this information.

**Step 2** State what is to be proven.

**Step 3** Create a deductive argument by forming a logical chain of statements linking the given information to what you are trying to prove.

**Step 4** Justify each statement with a reason. Reasons include definitions, algebraic properties, and theorems.

**Step 5** State what it is you have proven.

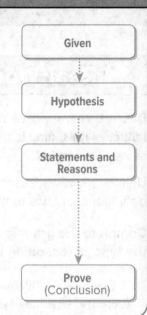

Given

↓

Hypothesis

↓

Statements and Reasons

↓

Prove (Conclusion)

*Work Zone*

A **proof** is a logical argument where each statement is justified by a reason. A **paragraph proof**, also called an **informal proof**, involves writing a paragraph to explain why a conjecture is true. In Example 1 below, you will use the algebraic property of substitution and the geometric relationship between vertical angles.

## Example

 Tutor

**1.** The diamondback rattlesnake has a diamond pattern on its back. An enlargement of the skin is shown. If $m\angle 1 = m\angle 4$, write a paragraph proof to show that $m\angle 2 = m\angle 3$.

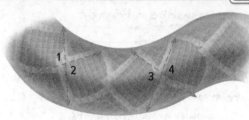

**Given:** $m\angle 1 = m\angle 4$

**Prove:** $m\angle 2 = m\angle 3$

**Proof:** $m\angle 1 = m\angle 2$ because they are vertical angles. Since $m\angle 1 = m\angle 4$, $m\angle 2 = m\angle 4$ by substitution. $m\angle 4 = m\angle 3$ because they are vertical angles. Since $m\angle 2 = m\angle 4$, then $m\angle 2 = m\angle 3$ also by substitution. Therefore, $m\angle 2 = m\angle 3$.

**Proofs**
Always end your proof with a statement that describes what you proved.

## Got It? Do this problem to find out.

a. Refer to the diagram shown. $AR = CR$ and $DR = BR$. Write a paragraph proof to show that $AR + DR = CR + BR$.

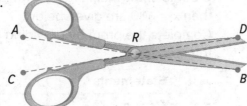

**Segment Notation**

$AR$ is read as the measure of line segment $AR$

**Given:** $AR =$ _____ and

$DR =$ _____ .

**Prove:** _____ $= CR + BR$.

**Proof:** You know that $AR = CR$ and $DR = BR$.

$AR + DR = CR + DR$ by the _____ Property of

Equality. So, $AR + DR = CR + BR$ by _____ .

# Two-Column Proofs

A **two-column proof** or **formal proof** contains *statements* and *reasons* organized in two columns. Once a statement or conjecture has been proven, it is called a **theorem**, and it can be used as a reason to justify statements in other proofs.

# Example

Tutor

**2.** Write a two-column proof to show that if two angles are vertical angles, then they have the same measure.

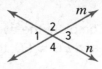

**Given:** lines $m$ and $n$ intersect; $\angle 1$ and $\angle 3$ are vertical angles

**Prove:** $m\angle 1 = m\angle 3$

| | Statements | Reasons |
|---|---|---|
| a. | lines $m$ and $n$ intersect; $\angle 1$ and $\angle 3$ are vertical angles. | Given |
| b. | $\angle 1$ and $\angle 2$ are a linear pair and $\angle 3$ and $\angle 2$ are a linear pair. | Definition of linear pair |
| c. | $m\angle 1 + m\angle 2 = 180°$ $m\angle 3 + m\angle 2 = 180°$ | Definition of supplementary angles |
| d. | $m\angle 1 + m\angle 2 = m\angle 3 + m\angle 2$ | Substitution |
| e. | $m\angle 1 = m\angle 3$ | Subtraction Property of Equality |

**Linear Pair**

A linear pair of angles is a pair of adjacent angles formed by intersecting lines.

> **Got It?** Do this problem to find out.

**b.** The statements for a two-column proof to show that if $m\angle Y = m\angle Z$, then $x = 100$ are given below. Complete the proof by providing the reasons.

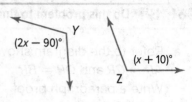

| Statements | Reasons |
|---|---|
| **a.** $m\angle Y = m\angle Z$, $m\angle Y = 2x - 90$, $m\angle Z = x + 10$ | |
| **b.** $2x - 90 = x + 10$ | |
| **c.** $x - 90 = 10$ | |
| **d.** $x = 100$ | |

# Guided Practice

1. Use the figure to complete the paragraph proof. (Example 1)

**Given:** $m\angle 1 = m\angle 2$, $\angle 1$ and $\angle 2$ are supplementary.

**Prove:** $\angle 1$ and $\angle 2$ are right angles.

**Proof:** $m\angle 1 + m\angle 2 =$ _____ since they are supplementary angles. Since $m\angle 1 = m\angle 2$, then $m\angle 1 + m\angle 1 = 180°$ by _____. Solving the equation gives $m\angle 1 =$ _____. Since $m\angle 1 = m\angle 2$, then $m\angle 2$ is also _____. Therefore, $\angle 1$ and $\angle 2$ are right angles.

2. Refer to the figure above. Complete the two-column proof to show that if $EG = 3x - 1$, $ED = 2x + 4$, and $EG = ED$, then $x = 5$. (Example 2)

| Statements | Reasons |
|---|---|
| **a.** $EG = 3x - 1$, $ED = 2x + 4$, $EG = ED$ | |
| **b.** $3x - 1 = 2x + 4$ | |
| **c.** $x - 1 = 4$ | |
| **d.** $x = 5$ | |

**Rate Yourself!**

Are you ready to move on?
Shade the section that applies.

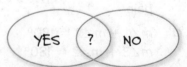

3. **Building on the Essential Question** How is deductive reasoning used in algebra and geometry proofs?

_____

_____

For more help, go online to access a Personal Tutor.

# Independent Practice

Go online for Step-by-Step Solutions

**1** In the figure at the right, two lines intersect to form four angles. If $m\angle 7 = 9x$ and $m\angle 8 = 11x$, complete the paragraph proof to show that $x = 9$. (Example 1)

**Given:** Two intersecting lines with $m\angle 7 = 9x$ and $m\angle 8 = 11x$

**Prove:** $x = 9$

**Proof:** $\angle 7$ and $\angle 8$ form a _____ angle so they are

_____ angles. So, $m\angle 7 + m\angle 8 =$ _____ , by the definition of supplementary angles. By substitution,

_____ $+ 11x = 180$. So, $x =$ _____ by the Division Property of Equality.

2. **MP Construct an Argument** Four towns lie on a straight road. Boyd is midway between Acton and Carson. Carson is midway between Boyd and Delta. Write a paragraph proof to show the distance from Acton to Boyd is the same as the distance from Carson to Delta. (Example 1)

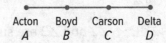

**Given:** $B$ is the midpoint of $\overline{AC}$ and $C$ is the midpoint of $\overline{BD}$.

**Prove:** $AB = CD$.

**Proof:** By the definition of midpoint, _____ $= BC$ and _____ $= CD$.

Therefore, $AB = CD$ by _____.

3. **MP Construct an Argument** Complete the two-column proof to show that if $\angle 1$ and $\angle 2$ are supplementary and $m\angle 1 = m\angle 2$, then $\angle 1$ and $\angle 2$ are right angles. (Example 2)

**Given:** $\angle 1$ and $\angle 2$ are supplementary; $m\angle 1 = m\angle 2$

**Prove:** $\angle 1$ and $\angle 2$ are right angles

| Statements | Reasons |
|---|---|
| a. $\angle 1$ and $\angle 2$ are supplementary; $m\angle 1 = m\angle 2$ | |
| b. $m\angle 1 + m\angle 2 = 180°$ | |
| c. $m\angle 1 + m\angle 1 = 180°$ | |
| d. $2(m\angle 1) = 180°$ | |
| e. $m\angle 1 = 90°$ | |
| f. $m\angle 2 = 90°$ | $m\angle 1 = m\angle 2$ (Given) |
| g. $\angle 1$ and $\angle 2$ are right angles. | |

4. **MP Construct an Argument** Complete the two-column proof to show that when two parallel lines are cut by a transversal, consecutive interior angles are supplementary.

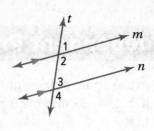

**Given:** parallel lines $m$ and $n$ cut by transversal $t$

**Prove:** $\angle 2$ and $\angle 3$ are supplementary.

| Statements | Reasons |
|---|---|
| a. _____ | Given |
| b. $\angle 1$ and $\angle 2$ form a straight angle. | _____ |
| c. _____ | Definition of supplementary angles |
| d. $m\angle 1 = m\angle 3$ | _____ |
| e. _____ | Substitution |
| f. $\angle 2$ and $\angle 3$ are supplementary angles | |

## H.O.T. Problems  Higher Order Thinking

5. **MP Reason Abstractly** Describe the theorem or definition you could use to find the measure of $\angle 2$.

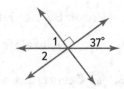

_____

_____

6. **MP Persevere with Problems** Given the right triangles shown, use inductive reasoning to make a conjecture about the sum of the measures of the two acute angles of any right triangle.

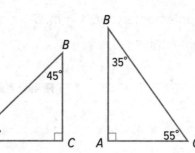

_____

_____

_____

7. **MP Reason Inductively** In the diagram, $m\angle CFE = 90°$ and $m\angle AFB = m\angle CFD$. Which of the following conclusions does *not* have to be true? _____

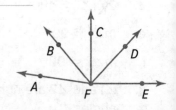

I  $m\angle AFB + m\angle DFE = 90°$    III  $m\angle CFD = m\angle AFB$

II  $\overline{BF}$ divides $\angle AFD$ in half    IV  $\angle CFE$ is a right angle.

# Extra Practice

8. **MP Construct an Argument** In the figure at the right, $AE = DB$ and C is the midpoint of $\overline{AE}$ and $\overline{DB}$. Complete the proof to show that $AC = CB$.

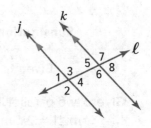

**Given:** $AE = DB$ and C is the midpoint of $\overline{AE}$ and $\overline{DB}$.

**Prove:** $AC = CB$

**Proof:** Since C is the midpoint of _____ and _____,

$$AC = CE = \frac{1}{2} \underline{\hspace{1cm}} \text{ and } DC = CB = \frac{1}{2} \underline{\hspace{1cm}} \text{ by the}$$

definition of midpoint. We are given $AE = DB$. By the _____

Property of Equality, $\frac{1}{2} AE = \frac{1}{2} DB$. So, by _____, $AC = CB$.

9. **MP Construct an Argument** Refer to the figure at the right. Complete the two-column proof to show if $m\angle 3 = 2x - 15$ and $m\angle 6 = x + 55$, then $x = 70$.

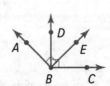

**Given:** $j \parallel k$, transversal $\ell$; $m\angle 3 = 2x - 15$, $m\angle 6 = x + 55$

**Prove:** $x = 70$

| Statements | Reasons |
|---|---|
| a.  $j \parallel k$, transversal $\ell$; $m\angle 3 = 2x - 15$, $m\angle 6 = x + 55$ | |
| b.  $m\angle 3 = m\angle 6$ | |
| c.  $2x - 15 = x + 55$ | |
| d.  $x - 15 = 55$ | |
| e.  $x = 70$ | |

10. **MP Construct an Argument** Refer to the figure at the right. Complete the two-column proof to show if $\angle ABE$ and $\angle DBC$ are right angles, then $m\angle ABD = m\angle EBC$.

**Given:** $\angle ABE$ and $\angle DBC$ are right angles.

**Prove:** $m\angle ABD = m\angle EBC$

| Statements | Reasons |
|---|---|
| a.  $\angle ABE$ and $\angle DBC$ are right angles. | |
| b.  $m\angle ABE = 90$ and $m\angle DBC = 90$ | |
| c.  $m\angle ABD + m\angle DBE = 90$ <br> $m\angle DBE + m\angle EBC = 90$ | |
| d.  $m\angle ABD + m\angle DBE = m\angle DBE + m\angle EBC$ | |
| e.  $m\angle ABD = m\angle EBC$ | |

**11.** In the diagram shown, $\overline{AE}$ intersects $\overline{DB}$ at $C$.

Determine if each of the following conclusions will always be true. Select yes or no.

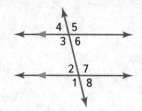

**a.** $m\angle ACD = m\angle BCE$ ☐ Yes ☐ No

**b.** $\angle ACD$ and $\angle ECD$ form a linear pair. ☐ Yes ☐ No

**c.** $\angle DCE$ and $\angle ACB$ are vertical angles. ☐ Yes ☐ No

**d.** $\angle ACB$ and $\angle BCE$ are complementary angles. ☐ Yes ☐ No

**12.** Select the appropriate reason for each statement of the geometric proof below.

| Substitution | Division Property of Equality | Vertical angles have equal measures. |
|---|---|---|
| Given | Alternate interior angles have equal measures. | Corresponding angles have equal measures. |

**Given:** two parallel lines cut by a transversal,
$m\angle 1 = 2x$, $m\angle 3 = 94$

**Prove:** $x = 47$

**Proof:**

| Statements | Reasons |
|---|---|
| **a.** $m\angle 1 = 2x$, $m\angle 3 = 94$ | |
| **b.** $m\angle 1 = m\angle 3$ | |
| **c.** $2x = 94$ | |
| **d.** $x = 47$ | |

Refer to the diagram. Identify each pair of angles as *adjacent*, *vertical*, or *neither*. 7.G.1

**13.** $\angle A$ and $\angle B$ _____

**14.** $\angle A$ and $\angle C$ _____

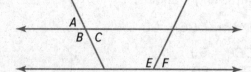

**15.** $\angle C$ and $\angle E$ _____

**16.** $\angle E$ and $\angle F$ _____

# Inquiry Lab
## Triangles

 **Inquiry** WHAT is the relationship among the measures of the angles of a triangle?

**CCSS** Content Standards 8.G.5

**MP** Mathematical Practices 1, 3

Lamont has a metal bracket that is in the shape of an angle that attaches a bag to the frame of a bike. The angle of the bracket measures 35°. Lamont wonders if it will fit into the frame of the bike by the handlebars.

## Hands-On Activity

Triangle means *three angles*. In this Activity you will explore how the three angles of a triangle are related.

**Step 1** On a separate piece of paper, draw a triangle like the one shown below.

**Step 2** Label the corners 1, 2, and 3. Then tear off each corner.

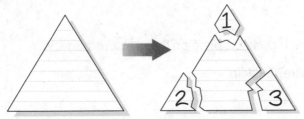

**Step 3** Rearrange the torn pieces so that the corners all meet at one point. Label the torn pieces with 1, 2, and 3.

What does each torn corner represent?

_____

The point where these corners meet is the vertex of another angle. Classify this angle as *acute*, *right*, *obtuse*, or *straight*.

Explain. _____

# Investigate

Collaborate

**Work with a partner. Repeat Steps 1–3 of the Activity on the previous page for each of the following triangles. Draw or tape your results in the space provided.**

1.

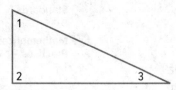

2.

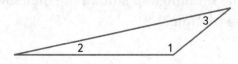

Show your work.

# Analyze and Reflect

Collaborate

3. **MP** **Reason Inductively** What is the sum of the measures of the angles for each

   of your triangles? _____

   Verify your conjecture below by measuring each angle using a protractor.

   Exercise 1: $m\angle 1 + m\angle 2 + m\angle 3 =$ _____

   Exercise 2: $m\angle 1 + m\angle 2 + m\angle 3 =$ _____

4. **MP** **Justify Conclusions** Refer to the bicycle problem on the previous page. Will

   the bracket fit exactly into Lamont's bike? Explain. _____

# Create

On Your Own

5. **MP** **Use Math Tools** Find a real-world example of a triangle. Measure the angles
   of the triangle. What is the sum of the measures of the angles? Does your answer

   support your findings in this Inquiry Lab? Explain. _____

   _____

   _____

   _____

6. **Inquiry** WHAT is the relationship among the measures of the angles of a

   triangle? _____

# Angles of Triangles

## Real-World Link

**STEM** Caroline and Emily are building a bridge out of toothpicks for a science competition. Emily thinks the sides should be constructed using triangles. Use the activity to find the sum of the measures of the angles in a triangle.

Collaborate

Lines *m* and *n* are parallel. Lines *p* and *r* are transversals that intersect at point *A*.

1. What is true about the measures of ∠1 and ∠2? Explain.

_____

_____

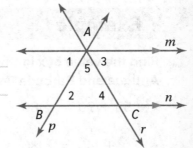

2. What is true about the measures of ∠3 and ∠4? Explain.

_____

_____

3. What kind of angle is formed by ∠1, ∠5, and ∠3? Write an equation representing the relationship between the 3 angles.

_____

4. Use the information from Exercises 1, 2, and 3 to draw a conclusion about the sum of the measures of the angles of △*ABC*. Explain your reasoning.

_____

_____

_____

**Essential Question**

HOW can algebraic concepts be applied to geometry?

Vocab
**Vocabulary**

triangle
interior angle
exterior angle
remote interior angles

**CCSS**    **Common Core State Standards**

**Content Standards**
8.G.5

  **Mathematical Practices**
1, 2, 3, 4

Which **MP** **Mathematical Practices** did you use?
Shade the circle(s) that applies.

① Persevere with Problems

② Reason Abstractly

③ Construct an Argument

④ Model with Mathematics

⑤ Use Math Tools

⑥ Attend to Precision

⑦ Make Use of Structure

⑧ Use Repeated Reasoning

## Angle Sum of a Triangle

**Words** The sum of the measures of the interior angles of a triangle is 180°.

**Model**

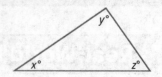

**Symbols** $x + y + z = 180°$

A **triangle** is formed by three line segments that intersect only at their endpoints. A point where the segments intersect is a vertex. The angle formed by the segments that lies inside the triangle is an **interior angle.**

 ## Example

 Tutor

**1.** **Find the value of x in the Antigua and Barbuda flag.**

$$x + 55 + 90 = 180 \quad \text{Write the equation.}$$

$$x + 145 = 180 \quad \text{Simplify.}$$

$$\underline{-145 = -145} \quad \text{Subtract.}$$

$$x = 35 \quad \text{Simplify.}$$

The value of x is 35.

**Got It?** Do this problem to find out.

a. _____

**a.** In △XYZ, if $m\angle X = 72°$ and $m\angle Y = 74°$, what is $m\angle Z$?

## Example

 Tutor

**2.** **The measures of the angles of △ABC are in the ratio 1:4:5. What are the measures of the angles?**

Let x represent the measure of angle A.

Then 4x and 5x represent angle B and angle C.

$$x + 4x + 5x = 180 \quad \text{Write the equation.}$$

$$10x = 180 \quad \text{Collect like terms.}$$

$$x = 18 \quad \text{Division Property of Equality}$$

Since x = 18, 4x = 4(18) or 72, and 5x = 5(18) or 90.
The measures of the angles are 18°, 72°, and 90°.

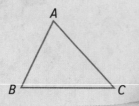

**Segments**

$\overline{AB}$ is read as segment AB. So the sides of the triangle below are $\overline{AB}$, $\overline{AC}$, and $\overline{BC}$.

## Got It? Do this problem to find out.

**b.** The measures of the angles of △LMN are in the ratio 2:4:6. What are the measures of the angles?

b. _____

# Exterior Angles of a Triangle

**Key Concept**

**Words** The measure of an exterior angle of a triangle is equal to the sum of the measures of its two remote interior angles.

**Model**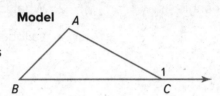

**Symbols** $m\angle A + m\angle B = m\angle 1$

In addition to its three interior angles, a triangle can have an **exterior angle** formed by one side of the triangle and the extension of the adjacent side. Each exterior angle of the triangle has two **remote interior angles** that are not adjacent to the exterior angle.

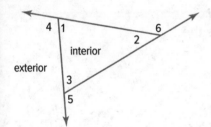

∠**4** is an exterior angle of the triangle. Its two remote interior angles are ∠**2** and ∠**3**.

$$m\angle 4 = m\angle 2 + m\angle 3$$

**STOP and Reflect**

Measure ∠2, ∠3, and ∠4 to verify that $m\angle 2 + m\angle 3 = m\angle 4$. Repeat the process for exterior angles 5 and 6. What is true about $m\angle 5$ and $m\angle 6$?

# Example

**Tutor**

**3.** Suppose $m\angle 4 = 135°$. Find the measure of ∠2.

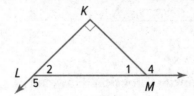

Angle 4 is an exterior angle. Its two remote interior angles are ∠2 and ∠LKM.

$$m\angle 2 + m\angle LKM = m\angle 4 \quad \text{Write the equation.}$$
$$x + 90° = 135° \quad m\angle 2 = x°, m\angle LKM = 90°, m\angle 4 = 135°$$
$$x = 45° \quad \text{Subtraction Property of Equality}$$

So, the $m\angle 2 = 45°$.

c. _____

**Got It?** Do this problem to find out.

c. Refer to the figure at the right.
Suppose $m\angle 5 = 147°$. Find $m\angle 1$.

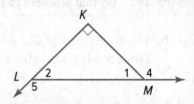

# Guided Practice

1. Find the value of $x$ in the triangle. (Example 1)

_____

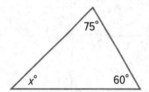

2. What is the value of $x$ in the sail of the

sailboat? (Example 1) _____

3. The measures of the angles of $\triangle LMN$ are in the ratio 1:2:5. What are the
measures of the angles? (Example 2)

_____

_____

4. Find the value of $x$ in the triangle. (Example 3) _____

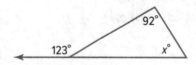

**Rate Yourself!**

Are you ready to move on?
Shade the section that applies.

I have a few questions.

I'm ready to move on.

I have a lot of questions.

5. **Building on the Essential Question** How can
you find the missing measure of an angle in a
triangle if you know the measure of two of the interior angles?

_____

_____

_____

_____

For more help, go online to
access a Personal Tutor.

Tutor

# Independent Practice

Go online for Step-by-Step Solutions

eHelp

**1.** The top of a kite is shown below. What is the value of *x*? (Example 1) _____

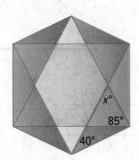

85°
40°
*x*°

**2.** A popular toy puzzle is shown below. What is the value of *x*? (Example 1) _____

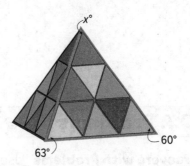

*x*°
63°
60°

**3.** The measures of the angles of △*RST* are in the ratio 2:4:9. What are the measures of the angles? (Example 2) _____

**4.** The measures of the angles of △*XYZ* are in the ratio 3:3:6. What are the measures of the angles? (Example 2) _____

**Find the value of *x* in each triangle.** (Example 3)

**5.** _____

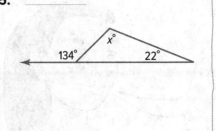

134°   *x*°   22°

**6.** _____

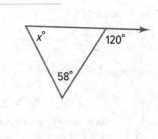

*x*°   120°
58°

**7.** _____

*x*°   125°          170°

**8.** In △*ABC* the measure of angle *A* is 2*x* + 3, the measure of angle *B* is 4*x* + 2, and the measure of angle *C* is 2*x* − 1. What are the measures of the angles? _____

**9** MP **Reason Abstractly** What is the measure of the third angle of a triangle if one angle measures 25° and the second angle measures 50°?

_____

**Find the measures of the angles in each triangle.**

**10.** _____
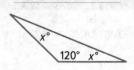

**11** _____
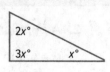

**12.** _____
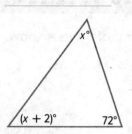

## H.O.T. Problems   Higher Order Thinking

**13.** **MP Persevere with Problems**  Use the figure at the right to informally prove that an exterior angle of a triangle is equal to the sum of its two remote interior angles.

**Given:** $\triangle ABC$; $\angle 1$ is an exterior angle.

**Prove:** $m\angle 1 = m\angle 2 + m\angle 3$

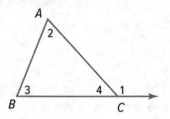

**Proof:** _____

_____

_____

_____

_____

_____

**14.** **MP Find the Error**  Alma is finding the measures of the angles in a triangle that have the ratio 1:3:5. Circle her mistake and correct it.

$$x + 3x + 5x = 180$$
$$8x = 180$$
$$x = 22.5$$

The angles measure 22.5°, 3(22.5) or 67.5°, and 5(22.5) or 122.5°.

_____

_____

_____

_____

_____

**15.** **MP Justify Conclusions**  Make a conjecture about the sum of the interior angles of a quadrilateral. Justify your reasoning.

_____

_____

_____

# Extra Practice

**Find the value of *x* in each triangle with the given angle measures.**

**16.**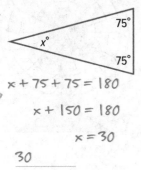

x + 75 + 75 = 180

x + 150 = 180

x = 30

30

**17.**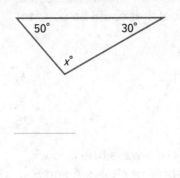

_____

_____

**18.**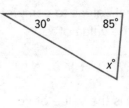

_____

_____

**19.** 70°, 60°, *x*° _____

**20.** *x*°, 60°, 25° _____

**21.** *x*°, 35°, 25° _____

**22.** The measures of the angles of △*DEF* are in the ratio 2:4:4. What are the measures of the angles?

_____

**23.** The measures of the angles of △*XYZ* are in the ratio 4:5:6. What are the measures of the angles?

_____

**Copy and Solve** Find the value of *x* in each triangle. Show your work on a separate sheet of paper.

**24.**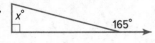

**25.**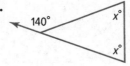

**26.**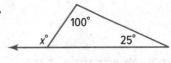

**27.** **MP** **Reason Inductively** Apply what you know about angles and lines to find the values of *x* and *y* in the figure at the right.

*x* = _____    *y* = _____

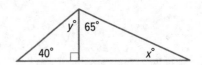

**28.** When viewed from the front, the base of an upright fan has a triangular face with the angle measures shown. Select the correct values to complete the model that could be used to find the value of *x*.

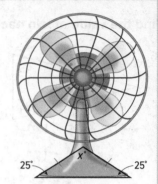

☐ + ☐ · ☐ = ☐

What is the value of *x*? ☐

| *x* | 65 |
|-----|-----|
| 2 | 90 |
| 25 | 180 |

**29.** Which of the following statements are always true about the relationship between the measures of angles *A* and *B* of the right triangle shown? Select all that apply.

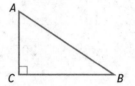

☐ They are equivalent.

☐ They are supplementary.

☐ They are acute.

☐ They are complementary.

## Common Core Spiral Review

**30.** The street maintenance vehicles for the city of Centerburg cannot safely make turns less than 70°. Should the proposed site of the new maintenance garage at the northeast corner of Park and Main be approved? Explain. **7.G.5**

_____

_____

**31.** ∠*A* and ∠*B* are complementary, and the measure of ∠*A* is 39°. What is

the measure of ∠*B*? **7.G.5** _____

**Solve each equation.** **6.EE.7**

**32.** $x + 72 + 63 + 120 = 360$

**33.** $90 + 90 + (2x + 4) + (3x - 29) = 360$

# Polygons and Angles

## Vocabulary Start-Up

A **polygon** is a simple closed figure formed by three or more line segments. The segments intersect only at their endpoints.

A map of the United States is shown. List the states that are in the shape of a polygon. Then list the number of segments that form the polygon.

| State | Number of Segments |
|-------|--------------------|
|       |                    |
|       |                    |
|       |                    |
|       |                    |

**Essential Question**

HOW can algebraic concepts be applied to geometry?

**Vocabulary**

polygon
equiangular
regular polygon

**CCSS** **Common Core State Standards**

**Content Standards**
Extension of 8.G.5

**MP** **Mathematical Practices**
1, 3, 4

Which **MP Mathematical Practices** did you use?
Shade the circle(s) that applies.

① Persevere with Problems
② Reason Abstractly
③ Construct an Argument
④ Model with Mathematics
⑤ Use Math Tools
⑥ Attend to Precision
⑦ Make Use of Structure
⑧ Use Repeated Reasoning

# Interior Angle Sum of a Polygon

**Words**  The sum of the measures of the interior angles of a polygon is $(n - 2)180$, where $n$ represents the number of sides.

**Symbols**  $S = (n - 2)180$

You can use the sum of the angle measures of a triangle to find the sum of the interior angle measures of various polygons. A polygon that is equilateral (all sides are the same length) and **equiangular** (all angles have the same measure) is called a **regular polygon**.

| Number of Sides | Sketch of Figure | Number of Triangles | Sum of Angle Measures |
|---|---|---|---|
| 3 | | 1 | $1(180°) = 180°$ |
| 4 | | 2 | $2(180°) = 360°$ |
| 5 | | 3 | $3(180°) = 540°$ |
| 6 | | 4 | $4(180°) = 720°$ |

**Everyday Use**
Deca- a prefix meaning ten, as in decade

**Math Use**
Decagon a polygon with ten sides

## Example

Tutor

1. **Find the sum of the measures of the interior angles of a decagon.**

   $S = (n - 2)\ 180$        Write an equation.

   $S = (10 - 2)\ 180$       A decagon has 10 sides. Replace $n$ with 10.

   $S = (8)180$ or $1,440$   Simplify.

   The sum of the measures of the interior angles of a decagon is $1,440°$.

Show your work.

**Got It?**  **Do these problems to find out.**

**Find the sum of the interior angle measures of each polygon.**

   **a.** hexagon          **b.** octagon          **c.** 15-gon

a. _____

b. _____

c. _____

## Example

Tutor

**2.** Each chamber of a bee honeycomb is a regular hexagon. Find the measure of an interior angle of a regular hexagon.

**Step 1** Find the sum of the measures of the angles.

$S = (n - 2)180$    Write an equation.

$S = (6 - 2)180$    Replace $n$ with 6.

$S = (4)180$ or 720    Simplify.

The sum of the measures of the interior angles is 720°.

**Step 2** Divide 720 by 6, the number of interior angles, to find the measure of one interior angle. So, the measure of one interior angle of a regular hexagon is 720° ÷ 6 or 120°.

Show your work.

d. _____

**Got It?** Do these problems to find out.

Find the measure of one interior angle in each regular polygon. Round to the nearest tenth if necessary.

   **d.** octagon           **e.** heptagon           **f.** 20-gon

e. _____

f. _____

# Exterior Angles of a Polygon

Key Concept

**Words**     In a polygon, the sum of the measures of the exterior angles, one at each vertex, is 360°.

**Model**

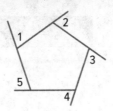

**Symbols**   $m\angle 1 + m\angle 2 + m\angle 3 + m\angle 4 + m\angle 5 = 360°$

Regardless of the number of sides in a polygon, the sum of the exterior angle measures is equal to 360°.

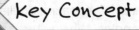

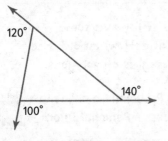

$120 + 100 + 140 = 360°$          $105 + 110 + 105 + 40 = 360°$

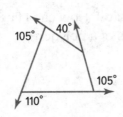
**STOP and Reflect**

Draw another quadrilateral and a pentagon. Extend the sides to show the exterior angles. Then find the sum of each figure's exterior angle measures.

## Example

**3.** Find the measure of an exterior angle in a regular hexagon.

Let $x$ represent the measure of each exterior angle.

$$6x = 360 \qquad \text{Write an equation. A hexagon has 6 exterior angles.}$$
$$x = 60 \qquad \text{Division Property of Equality}$$

So, each exterior angle of a regular hexagon measures 60°.

*Show your work.*

g. _____

h. _____

i. _____

**Got It?** Do these problems to find out.

Find the measure of an exterior angle of each regular polygon.

**g.** triangle  **h.** quadrilateral  **i.** octagon

## Guided Practice

Find the sum of the interior angle measures of each polygon. (Example 1)

**1.** quadrilateral _____

**2.** nonagon _____

**3.** 12-gon _____

*Show your work.*

**4.** The quilt pattern shown is made of repeating equilateral triangles. What is the measure of one interior angle of an equilateral triangle? (Example 2)

_____

**5.** Find the measure of an exterior angle of a regular pentagon. (Example 3) _____

**6.** ⓔ **Building on the Essential Question** How can I find the sum of the interior angle measures of a polygon?

_____

_____

_____

### Rate Yourself!

☐ I understand how to find the sum of the interior angle measures of a polygon.

▷▷ Great! You're ready to move on!

☐ I still have some questions about the angles of polygons.

📖 No Problem! Go online to access a Personal Tutor. **Tutor**

Name _____

# Independent Practice

Go online for Step-by-Step Solutions

**Find the sum of the interior angle measures of each polygon.** (Example 1)

**1.** pentagon _____

**2.** 11-gon _____

**3** 13-gon _____

**4.** The soccer ball at the right consists of repeating regular pentagons and hexagons. Find the measure of one interior angle of a pentagon.

(Example 2) _____

**Find the measure of an exterior angle of each regular polygon.** (Example 3)

**5** decagon _____

**6.** 20-gon _____

**7.** 15-gon _____

**A tessellation is a repetitive pattern of polygons that fit together without overlapping and without gaps between them. For each tessellation, find the measure of each angle at the circled vertex. Then find the sum of the angles.**

**8.**

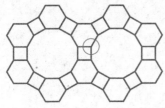

_____

**9.**

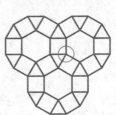

_____

**Find the value of x in each figure.**

**10.** _____

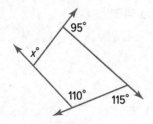

95°
x°
110°    115°

**11.** _____

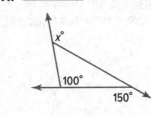

x°
100°
150°

12. **MP** **Model with Mathematics** Refer to the graphic novel frame below. Find the measures of the two missing angles using properties of quadrilaterals and parallel lines. _____

## H.O.T. Problems Higher Order Thinking

13. **MP** **Persevere with Problems** How many sides does a regular polygon have if the measure of an interior angle is 160°? Justify your answer.

14. **MP** **Reason Inductively** If the number of sides of a polygon increases by 1, what happens to the sum of the measures of the interior angles?

_____

_____

15. **MP** **Reason Inductively** Devon drew a regular polygon and measured one of its interior angles. Explain why it is impossible for his angle measure to be 145°.

_____

_____

_____

# Extra Practice

**Find the sum of the interior angle measures of each polygon.**

**16.** heptagon _900°_

$s = (n - 2)180$

$s = (7 - 2)180$

$s = 5 \cdot 180$

$s = 900$

**17.** 14-gon _____

**18.** 24-gon _____

**Find the measure of one interior angle in each regular polygon. Round to the nearest tenth if necessary.**

**19.** nonagon _____

**20.** decagon _____

**21.** 19-gon _____

**22.** 16-gon _____

**Find the measure of an exterior angle of each regular polygon.**

**23.** nonagon _____

**24.** 12-gon _____

**25.** 18-gon _____

**26.** The surface of the dome of Spaceship Earth in Orlando consists of repeating equilateral triangles as shown. Find the measure of each angle in each outlined triangle. Then make a conjecture about the interior angle measures in equilateral triangles of different sizes.

_____

_____

_____

**27.** **MP** **Justify Conclusions** What is the sum of the interior angles of nonregular hexagons? Explain your reasoning to a classmate.

_____

_____

**28.** After the first two folds of an origami paper design, the paper is shaped like a square with two isosceles triangles removed from two adjacent corners.

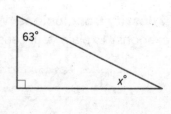

Angles *AED* and *BCD* are congruent. Select the correct values to complete the model below to find the measure of angle *AED*.

| *x* | 2 | 3 | 45 | 90 |
|-----|-----|-----|-----|-----|
| 180 | 360 | 540 | 720 | |

☐ · ☐ + ☐ · ☐ = ☐

What is *m∠AED*? ☐

**29.** Fill in each box to make each statement true.

**a.** The sum of the interior angle measures of a quadrilateral is ☐.

**b.** The sum of the interior angle measures of a(n) ☐ is 720°.

**c.** The sum of the interior angle measures of an octagon is ☐.

**d.** The sum of the interior angle measures of a(n) ☐ is 1,620°.

Classify each pair of angles as *complementary*, *supplementary*, or *neither*. 7.G.5

**30.** angle 1: 35° _____
angle 2: 55°

Show your work.

**31.** angle 1: 62° _____
angle 1: 108°

Find the value of *x* in each triangle. 8.G.5

**32.** _____

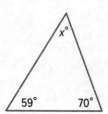

**33.** _____

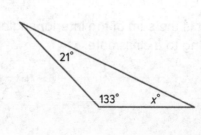

**34.** _____

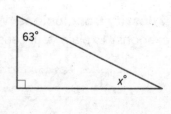

**Problem-Solving Investigation**
# Look for a Pattern

## Case #1 Spider Web

An activity in a low ropes course creates the inside of a spider web using string. The group members form a polygon. The strings stretch from each person to every nonadjacent member of the figure. Chen's group has 20 members.

*How many strings will Chen hold in the web?*

**Content Standards**
8.G.5
**MP** Mathematical Practices
1, 3, 8

 **Understand** *What are the facts?*
- There are 20 group members that form a polygon.
- A string stretches from each person to every nonadjacent group member.

 **Plan** *What is your strategy to solve this problem?*
Drawing a 20-gon would be difficult. Begin with a group of four members and look for a pattern. Then make a table to find the pattern.

 **Solve** *How can you apply the strategy?*
Draw figures using four, five, and six members. Draw the diagonals from one member to show number of strings. Some figures are drawn for you.

4 members          5 members          6 members

| Number of Members | 4 | 5 | 6 | 7 | 8 | 9 |
|---|---|---|---|---|---|---|
| Number of Strings | | | | | | |

How many strings will Chen hold? _____

 **Check** *Does the answer make sense?*
Draw a 20-gon and count the number of diagonals from one vertex.

## Analyze the Strategy  Tutor

**MP** **Identify Repeated Reasoning** How would the pattern change if Chen was looking for the total number of strings from every person in the web?

_____

_____

## Case #2  Follow the Bouncing Ball

A ball was dropped from a height of 27 inches. After the first, second, and third bounces, the heights were 18 inches, 12 inches, and 8 inches, respectively.

After which bounce will the height of the ball be less than 3 inches?

## Understand

**Read the problem. What are you being asked to find?**

I need to find _____.

**Underline key words and values. What information do you know?**

The ball is dropped from ☐ inches. The first bounce is ☐ inches high, the second bounce is ☐ inches high, and the third bounce is ☐ inches high.

## Plan

**Choose a problem-solving strategy.**

I will use the _____ strategy.

## Solve

**Use your problem-solving strategy to solve the problem.**

+1  +1

| Bounce | | | | | | | |
|---|---|---|---|---|---|---|---|
| Height (in.) | | | | | | | |

$\times \frac{2}{3}$  $\times \frac{2}{3}$

So, _____

_____.

## Check

**Use information from the problem to check your answer.**

**Work with a small group to solve the following cases.**
**Show your work on a separate piece of paper.**

Collaborate

## Case #3 Geometry

Right triangles are arranged as shown. The sum of the measures of the angles in the first figure is 360°.

What is the sum of the measures of the angles in the fifth figure?

## Case #4 Seating

A theater has 12 seats in the first row, 17 seats in the second row, 22 seats in the third row, and so on.

How many seats are in the eighth row? the nth row?

## Case #5 Mental Math Tricks

Study the pattern.

Without doing the multiplication, what is the answer to
$1,111,111 \times 1,111,111$?

$$1 \times 1 = 1$$
$$11 \times 11 = 121$$
$$111 \times 111 = 12,321$$
$$1111 \times 1111 = 1,234,321$$

## Case #6 Time

Use any strategy!

Carlos and his friends are going out to bowl, eat dinner, and see a movie. The movie starts at 8:10 P.M. and they want to arrive 20 minutes before it starts. They will bowl for one hour and dinner will take 1 hour and 15 minutes. Travel time is 20 minutes to the bowling alley, 45 minutes to the restaurant, and 10 minutes to the theater.

At what time should they plan to leave Carlos' house?

# Mid-Chapter Check

## Vocabulary Check

1. Name a pair of angles for each of the following. (Lesson 1)

   a. corresponding angles _____

   b. alternate interior angles _____

   c. vertical angles _____

   d. alternate exterior angles _____

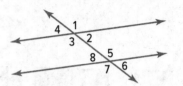

2. List two attributes of regular polygons. (Lesson 4)

   _____

   _____

## Skills Check and Problem Solving

**Refer to the figure at the right. Find the missing measure of each angle.** (Lessons 1 and 3)

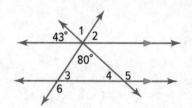

3. $m\angle 1 =$ _____

4. $m\angle 2 =$ _____

5. $m\angle 3 =$ _____

6. $m\angle 4 =$ _____

7. $m\angle 5 =$ _____

8. $m\angle 6 =$ _____

9. The building for the headquarters for the United States Department of Defense is in the shape of a regular polygon with five sides. What is the measure of one of the interior angles of the building? (Lesson 4) _____

10. **MP Use Math Tools** In the figure, line $a$ is parallel to line $b$. Which of the following are equal to the measure of $\angle T$? (Lesson 1) _____

    I   the supplement of $\angle S$

    II  the complement of $\angle X$

    III the angle adjacent to $\angle Z$

    IV  angle corresponding to $\angle R$

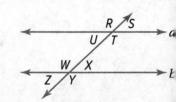

# Inquiry Lab
## Right Triangle Relationships

 **Inquiry** WHAT is the relationship among the sides of a right triangle?

 Content Standards 8.G.6

 Mathematical Practices 1, 3, 4

Three square tents at the school carnival are situated as shown below. The backs of the orange tent and the green tent form a right angle. The back of the blue tent finishes the triangle.

## Hands-On Activity

Using grid paper can help to investigate the relationship between the sides of a right triangle.

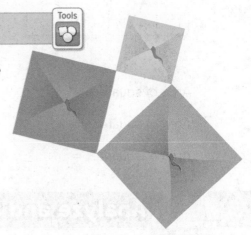

**Step 1** In each figure below, the sides of three squares form a right triangle.

Triangle 1    Triangle 2    Triangle 3

**Step 2** Find the area of each square that is attached to the triangle. Record your results in the table below. The first one is already done for you. Use the figures at the right to help find the area of partial grids.

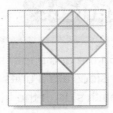

Area = $\frac{1}{2}$ cm²    Area = 1 cm²

| Triangle | Area of Green Square | Area of Blue Square | Area of Yellow Square |
|---|---|---|---|
| 1 | 1 | 1 | 2 |
| 2 | | | |
| 3 | | | |

What relationship exists among the areas of the three squares bordering each triangle? _____

_____

## Investigate

Work with a partner. Draw a right triangle different than those of the Activity on grid paper. Find the area of each square that is attached to the triangle.

**1.**

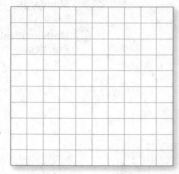

Show your work.

**2.**

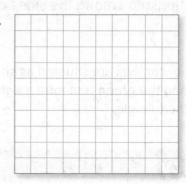

Area of square 1 = _____

Area of square 2 = _____

Area of square 3 = _____

Area of square 1 = _____

Area of square 2 = _____

Area of square 3 = _____

## Analyze and Reflect

**3.**  **MP Model with Mathematics** On the grid paper shown, draw a right triangle so the two shorter sides are 3 units and 4 units long. Draw squares attached to each side of the triangle.

What is the area of each square?

_____

What is the length of each side?

_____

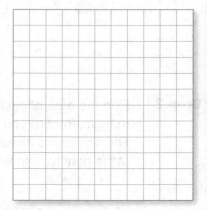

## Create

On Your Own

**4.** **MP Reason Inductively** Make a conjecture about the length of the longest side of a right triangle if the lengths of the two shorter sides are 6 centimeters and 8 centimeters.

_____

_____

**5.** **Inquiry** WHAT is the relationship among the sides of a right triangle?

_____

_____

# The Pythagorean Theorem

## Vocabulary Start-Up

A right triangle is a triangle with one right angle. The **legs** are the sides that form the right angle. The **hypotenuse** is the side opposite the right angle. It is the longest side of the triangle.

**Complete the graphic organizer. Label the legs and the hypotenuse.**

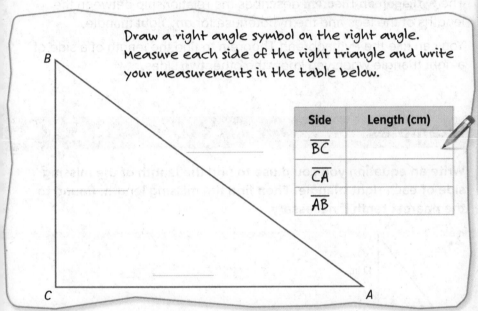

Draw a right angle symbol on the right angle. Measure each side of the right triangle and write your measurements in the table below.

| Side | Length (cm) |
|------|-------------|
| $\overline{BC}$ | |
| $\overline{CA}$ | |
| $\overline{AB}$ | |

 **Essential Question**

HOW can algebraic concepts be applied to geometry?

 **Vocabulary**

legs
hypotenuse
Pythagorean Theorem
converse

**Common Core State Standards**

**Content Standards**
8.G.7, 8.EE.2

**MP Mathematical Practices**
1, 3, 4, 5

##  Real-World Link

When viewed from the side, the shape of some wooden waterskiing ramps is a right triangle. Suppose the height of a ramp is 3 feet and the length of the base of the ramp is 4 feet. How long do you think the ramp will be? Explain your reasoning.

**Which MP Mathematical Practices did you use?**
**Shade the circle(s) that applies.**

① Persevere with Problems
② Reason Abstractly
③ Construct an Argument
④ Model with Mathematics
⑤ Use Math Tools
⑥ Attend to Precision
⑦ Make Use of Structure
⑧ Use Repeated Reasoning

# Pythagorean Theorem

**Words**  In a right triangle, the sum of the squares of the lengths of the legs is equal to the square of the length of the hypotenuse.

**Model**

**Symbols**  $a^2 + b^2 = c^2$

The **Pythagorean Theorem** describes the relationship between the lengths of the legs and the hypotenuse for *any* right triangle.

You can use the Pythagorean Theorem to find the length of a side of a right triangle when you know the other two sides.

## Examples

Tutor

Write an equation you could use to find the length of the missing side of each right triangle. Then find the missing length. Round to the nearest tenth if necessary.

**1.**

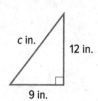

c in.  12 in.

9 in.

| | |
|---|---|
| $a^2 + b^2 = c^2$ | Pythagorean Theorem |
| $12^2 + 9^2 = c^2$ | Replace $a$ with 12 and $b$ with 9. |
| $144 + 81 = c^2$ | Evaluate $12^2$ and $9^2$. |
| $225 = c^2$ | Add 81 and 144. |
| $\pm\sqrt{225} = c$ | Definition of square root |
| $c = 15$ or $-15$ | Simplify. |

The equation has two solutions, 15 and −15. However, the length of a side must be positive. So, the hypotenuse is 15 inches long.

Check:  $a^2 + b^2 = c^2$

$12^2 + 9^2 \stackrel{?}{=} 15^2$

$144 + 81 \stackrel{?}{=} 225$

$225 = 225$ ✓

**Right Angle**

The symbol ⌐ indicates an angle with a measure of 90°.

**2.**

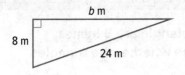

$$a^2 + b^2 = c^2 \qquad \text{Pythagorean Theorem}$$
$$8^2 + b^2 = 24^2 \qquad \text{Replace } a \text{ with 8 and } c \text{ with 24.}$$
$$64 + b^2 = 576 \qquad \text{Evaluate } 8^2 \text{ and } 24^2$$
$$64 - 64 + b^2 = 576 - 64 \qquad \text{Subtract 64 from each side.}$$
$$b^2 = 512 \qquad \text{Simplify.}$$
$$b = \pm\sqrt{512} \qquad \text{Definition of square root}$$
$$b \approx 22.6 \text{ or } -22.6 \qquad \text{Use a calculator.}$$

The length of side $b$ is about 22.6 meters.

**Check for Reasonableness** The hypotenuse is always the longest side in a right triangle. Since 22.6 is less than 24, the answer is reasonable.

**Got It?** Do these problems to find out.

**a.**
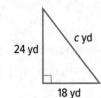

**b.**

**c.**
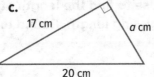

a. _____

b. _____

c. _____

# Converse of Pythagorean Theorem

**Key Concept**

If the sides of a triangle have lengths $a$, $b$, and $c$ units such that $a^2 + b^2 = c^2$, then the triangle is a right triangle.

If you reverse the parts of the Pythagorean Theorem, you have formed its **converse**.

**Statement:** If a **triangle is a right triangle**, then $a^2 + b^2 = c^2$.

**Converse:** If $a^2 + b^2 = c^2$, then the **triangle is a right triangle**.

The converse of the Pythagorean Theorem is also true.

## Example

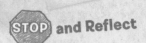

**3.** The measures of three sides of a triangle are 5 inches, 12 inches, and 13 inches. Determine whether the triangle is a right triangle.

$$a^2 + b^2 = c^2 \quad \text{Pythagorean Theorem}$$
$$5^2 + 12^2 \overset{?}{=} 13^2 \quad a = 5, b = 12, c = 13$$
$$25 + 144 \overset{?}{=} 169 \quad \text{Evaluate } 5^2, 12^2, \text{ and } 13^2.$$
$$169 = 169 \checkmark \quad \text{Simplify.}$$

The triangle is a right triangle.

**Got It?** Do these problems to find out.

Determine whether each triangle with sides of given lengths is a right triangle. Justify your answer.

**d.** 36 mi, 48 mi, 60 mi

**e.** 4 ft, 7 ft, 5 ft

**STOP and Reflect**

State three measures that could be the side measures of a right triangle. Justify your answer below.

d. _____

e. _____

# Guided Practice

Write an equation you could use to find the length of the missing side of each right triangle. Then find the missing length. Round to the nearest tenth if necessary. (Examples 1 and 2)

**1.** _____

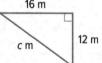

16 m, 12 m, c m

**2.** _____

100 mm, c mm, 200 mm

Determine whether each triangle with sides of given lengths is a right triangle. Justify your answer. (Example 3)

**3.** 5 in., 10 in., 12 in. _____

**4.** 9 m, 40 m, 41 m _____

**5.** **Building on the Essential Question** What is the relationship among the legs and the hypotenuse of a right triangle?

_____

**Rate Yourself!**

How confident are you about using the Pythagorean Theorem? Check the box that applies.

For more help, go online to access a Personal Tutor.

**FOLDABLES** Time to update your Foldable!

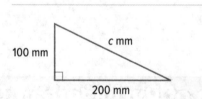

# Inquiry Lab

## Proofs About the Pythagorean Theorem

 **Inquiry** HOW can you prove the Pythagorean Theorem and its converse?

 **CCSS** Content Standards
8.G.6

 **MP** Mathematical Practices
1, 3, 7

The Pythagorean Theorem is named after a famous Greek mathematician Pythagoras who lived around 500 B.C. The properties of the theorem, however, were known by the ancient Egyptians, Babylonians, and Chinese. The following geometric proof is similar to a visual proof shown in a Chinese document written between 500 B.C. and 200 B.C.

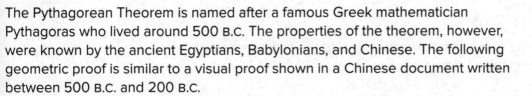

# Hands-On Activity 1

**Step 1** Draw and cut out 8 copies of a right triangle. Label each pair of legs *a* and *b*, and each hypotenuse *c*.

**Step 2** On a separate piece of paper arrange four of the triangles in a square as shown. Trace the figure formed by the hypotenuses.

The length of each side of the large square is $a + b$, so the area of the large square is $(a + b)^2$.

Is the figure formed by the hypotenuses a square? Explain.

_____

_____

Write an expression for the area of the inside square. _____

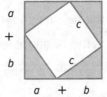

**Step 3** On the same paper, arrange the remaining triangles as shown. Draw the two figures shown by the dashed lines.

The length of each side of the large square is $a + b$, so the area of the large square is $(a + b)^2$.

Are the two figures represented by dashed line squares? Explain.

_____

_____

Write an expression for the area of the small square. ☐

Write an expression for the area of the large square. ☐

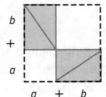

**Step 4**  Since the area of each of the two composite figures you created is $(a + b)^2$, the areas are equal. Use the space provided to draw each figure from Step 2 and Step 3. Place an equal sign between the two figures to show the two areas are equal.

**Step 5**  Remove the triangles from each side. Use the space provided to draw the remaining figures.

What property justifies removing the triangles from each side of the equation? _____

Write an algebraic equation that represents the relationship between the figures shown in Step 5. _____

Summarize the relationship among the sides of a right triangle measuring $a$ units, $b$ units, and $c$ units.

_____

_____

## Investigate

Collaborate

**Work with a partner.**

1. A legend states that the ancient Egyptians could create a right triangle by using a knotted rope. Research this on the Internet. Describe their technique in the space provided and draw a diagram to illustrate the technique.

Show your work.

_____

_____

_____

_____

# Hands-On Activity 2

The converse of the Pythagorean Theorem states if a triangle has side lengths, $a$, $b$, and $c$ units such that $a^2 + b^2 = c^2$, then the triangle is a right triangle. In this Activity, you will prove the converse of the Pythagorean Theorem by using a two-column proof.

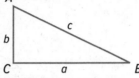

**Given:** $\triangle ABC$ such that $a^2 + b^2 = c^2$.

**Prove:** $\triangle ABC$ is a right triangle.

**Complete the proof with the correct reasons justifying each statement.**

| Statements | Reasons |
|---|---|
| **a.** Draw a right triangle *DEF* so that $\overline{DE}$ is $a$ units long and $\overline{DF}$ is $b$ units long. Label $\overline{FE}$ as $d$. | 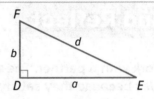 |
| **b.** Write an equation that describes the relationship between the side lengths of $\triangle DEF$. State the theorem that allows you to make that statement. | |
| **c.** $a^2 + b^2 = c^2$ | Given |
| **d.** If $a^2 + b^2 = c^2$ and $a^2 + b^2 = d^2$, then $d^2 = c^2$. | |
| **e.** If $d^2 = c^2$, then $d = c$. | |
| **f.** If $d = c$, then $FE = AB$. | |
| **g.** If $AC = FD$, $CB = DE$, and $AB = FE$, the two triangles are the same shape and size. | If three sides of a triangle are the same length as the corresponding sides of another triangle, the triangles are the same shape and size. |
| **h.** $m\angle C = m\angle D$ | Corresponding parts of the triangles with the same size and shape have the same measures. |
| **i.** $\angle C$ is a right angle. | |
| **j.** $\triangle ABC$ is a right triangle | |

So, if a triangle has side lengths, $a$, $b$, and $c$ units such that $a^2 + b^2 = c^2$, then the triangle is a right triangle.

## Investigate

Collaborate

**Work with a partner. Determine whether the following figures are right triangles. Justify your answer.**

**2.**

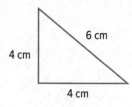

6 cm

4 cm

4 cm

**3.**

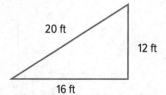

20 ft

12 ft

16 ft

Show your work.

_____    _____

## Analyze and Reflect

Collaborate

MP **Justify Conclusions** Work with a partner. The whole numbers 3, 4, and 5 are called *Pythagorean Triples* because they satisfy the Pythagorean Theorem. Determine if each of the following is a Pythagorean Triple. Explain your reasoning.

**4.** 7, 24, 25

_____

_____

**5.** 15, 20, 25

_____

_____

**6.** 9, 12, 16

_____

_____

## Create

On Your Own

**7.** MP **Identify Structure** In the Activity on page 409, you examined the relationship between the sides of a right triangle. Compare the process used in that Activity with the one on page 419. What kind of reasoning was used in each Activity?

_____

_____

_____

_____

_____

**8.** inquiry HOW can you prove the Pythagorean Theorem and its converse?

_____

_____

_____

# Use the Pythagorean Theorem

## Real-World Link

**Parasailing** In parasailing, a towrope is used to attach a parasailer to a boat. Refer to the diagram below for Exercises 1–4.

1. What type of triangle is formed by the horizontal distance, the vertical height, and the length of the towrope? Explain.

_____

_____

_____

2. Suppose the wind picks up and the parasailer rises to 50 feet and remains 72 feet behind the boat. Write an equation that will help you find how much towrope *c* the parasailer will need.

_____

3. Solve the equation to find the amount of rope the parasailer will

   need. Round to the nearest foot. ☐ ft

4. Suppose the towrope is 300 feet long and the parasailer is 200 feet above the water surface. Write an equation to find the horizontal distance *b* behind the boat.

_____

75 ft

21 ft

72 ft

---

**Which MP Mathematical Practices did you use?**
**Shade the circle(s) that applies.**

① Persevere with Problems      ⑤ Use Math Tools

② Reason Abstractly            ⑥ Attend to Precision

③ Construct an Argument        ⑦ Make Use of Structure

④ Model with Mathematics       ⑧ Use Repeated Reasoning

## Solve a Right Triangle

The Pythagorean Theorem can be used to solve a variety of problems. It is helpful to use a diagram to determine what part of the right triangle is unknown.

### Examples

**1.** **Write an equation that can be used to find the length of the ladder. Then solve. Round to the nearest tenth.**

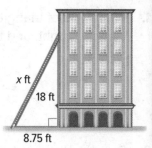

Notice that the distance from the building, the building itself, and the ladder form a right triangle. Use the Pythagorean Theorem.

$$a^2 + b^2 = c^2 \qquad \text{Pythagorean Theorem}$$
$$8.75^2 + 18^2 = c^2 \qquad \text{Replace } a \text{ with 8.75 and } b \text{ with 18.}$$
$$76.5625 + 324 = c^2 \qquad \text{Evaluate } 8.75^2 \text{ and } 18^2.$$
$$400.5625 = c^2 \qquad \text{Add 76.5625 and 324.}$$
$$\pm\sqrt{400.5625} = c \qquad \text{Definition of square root}$$
$$\pm 20.0 \approx c \qquad \text{Use a calculator.}$$

Since length cannot be negative, the ladder is about 20 feet long.

- - - - - - - - - - - - - - - - - - - - - - - - - - - - - -

**2.** **Write an equation that can be used to find the height of the plane. Then solve. Round to the nearest tenth.**

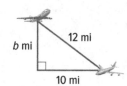

The distance between the planes is the hypotenuse of a right triangle. Use the Pythagorean Theorem.

$$a^2 + b^2 = c^2 \qquad \text{Pythagorean Theorem}$$
$$10^2 + b^2 = 12^2 \qquad \text{Replace } a \text{ with 10 and } c \text{ with 12.}$$
$$100 + b^2 = 144 \qquad \text{Evaluate } 10^2 \text{ and } 12^2.$$
$$b^2 = 44 \qquad \text{Subtraction Property of Equality}$$
$$b = \pm\sqrt{44} \qquad \text{Definition of square root.}$$
$$b \approx \pm 6.6 \qquad \text{Use a calculator.}$$

Since length cannot be negative, the height of the plane is about 6.6 miles.

## Got It? Do this problem to find out.

show your work.

**a.** Mr. Parsons wants to build a new banister for the staircase shown. If the *rise* of the stairs of a building is 5 feet and the *run* is 12 feet, what will be the length of the new banister?

a. _____

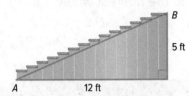

B

5 ft

A        12 ft

# The Pythagorean Theorem in Three-Dimensions

You can use the Pythagorean Theorem to find missing measures in three-dimensional figures.

Watch | Tutor

## Example

**3.** A 12-foot flagpole is placed in the center of a square area. To stabilize the pole, a wire will stretch from the top of the pole to each corner of the square. The flagpole is 7 feet from each corner of the square. What is the length of each wire? Round to the nearest tenth.

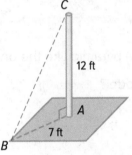

C

12 ft

A

7 ft

B

Draw right triangle *ABC*. You want to find the length of each wire or *BC*. This is the hypotenuse of a right triangle, so use the Pythagorean Theorem.

$$AB^2 + AC^2 = BC^2 \quad \text{Pythagorean Theorem}$$

$$7^2 + 12^2 = BC^2 \quad \text{Replace } AB \text{ with 7 and } AC \text{ with 12.}$$

$$49 + 144 = BC^2 \quad \text{Evaluate } 7^2 \text{ and } 12^2.$$

$$193 = BC^2 \quad \text{Simplify.}$$

$$\pm\sqrt{193} = BC \quad \text{Definition of square root.}$$

$$\pm 13.9 \approx BC \quad \text{Use a calculator.}$$

Since length cannot be negative, the length of the wire is about 13.9 feet.

**Got It?** Do this problem to find out.

b. _____

**b.** The top part of a circus tent is in the shape of a cone. The tent has a radius of 50 feet. The distance from the top of the tent to the edge is 61 feet. How tall is the top part of the tent? Round to the nearest whole number.

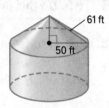

61 ft

50 ft

## Guided Practice

Check

Write an equation that can be used to answer the question. Then solve. Round to the nearest tenth if necessary. (Examples 1 and 2)

**1.** What is the height of the tent?

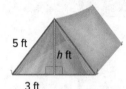

5 ft
*h* ft
3 ft

Show your work.

**2.** How high is the wheelchair ramp?

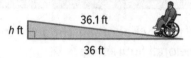

*h* ft
36.1 ft
36 ft

**3.** Merideth made a model of a pyramid like the one shown for history class.

What is the height of the model? (Example 3) _____

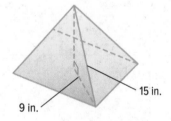

15 in.
9 in.

**4.** **Building on the Essential Question** How do you solve a right triangle?

_____

_____

_____

_____

**Rate Yourself!**

☐ I understand how to apply the Pythagorean Theorem.

▶▶ Great! You're ready to move on!

☐ I still have questions about how to apply the Pythagorean Theorem.

📖 No Problem! Go online to access a Personal Tutor.

Tutor

**FOLDABLES** Time to update your Foldable!

# Extra Practice

**12.** Write an equation to find how far the bird is from the boy. Then solve the equation. Round to the nearest tenth.

$70^2 + 20^2 = x^2; 72.8$ ft

$a = 70, b = 20,$ and $c = x$

$70^2 + 20^2 = x^2$

$4,900 + 400 = x^2$

$5,300 = x^2$

$\sqrt{5,300} = x^2$

$72.8 \approx x$

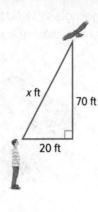

x ft · 70 ft · 20 ft

**13.** A party hat is in the shape of a cone with dimensions shown.

Find the height of the hat. Round to the nearest tenth. _____

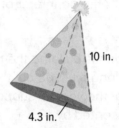

10 in. · 4.3 in.

**14.** Larry wants to go from his house to his grandmother's house. How much distance is saved if he takes Main Street instead of Market and Exchange?

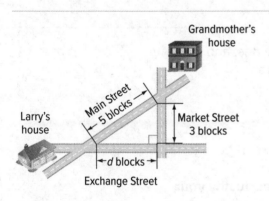

Larry's house · Main Street 5 blocks · Grandmother's house · Market Street 3 blocks · ←d blocks→ · Exchange Street

**15.** Suppose Greenville, Rock Hill, and Columbia form a right triangle. What is the distance from Columbia to Greenville?

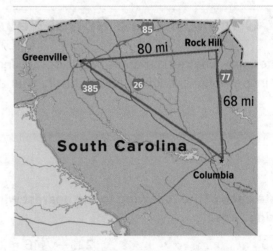

Greenville · 80 mi · Rock Hill · 68 mi · South Carolina · Columbia

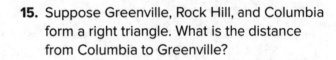

**MP** **Persevere with Problems** Find the missing measure in each figure below. Round to the nearest tenth if necessary.

**16.** _____

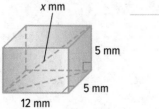

x mm · 5 mm · 5 mm · 12 mm

**17.** _____

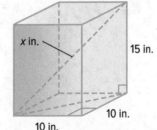

x in. · 15 in. · 10 in. · 10 in.

**18.** Shanise designed a stained glass window in the shape of a kite. Select the correct measures to label the dimensions of the window.

| 12 in. | 42 in. |
|--------|--------|
| 31 in. | 45 in. |
| 36 in. | 60 in. |
| 39 in. | |

15 in.

45 in.    27 in.

What is the perimeter of the window?

**19.** Brayden is building the model bridge shown. How long must he cut the piece of wood for one of the vertical support beams, represented by *x*?

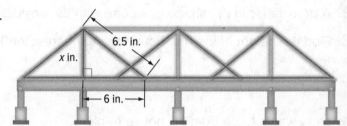
6.5 in.
*x* in.
6 in.

## Common Core Spiral Review

**20.** Determine whether a triangle with sides 20 inches, 48 inches, and 52 inches long is a right triangle. Justify your answer. **8.G.5**

**Estimate each of the following to the nearest whole number. Justify your reasoning. 8.NS.2**

**21.** $\sqrt{39} \approx$ _____

**22.** $-\sqrt{146} \approx$ _____

**23.** $\sqrt[3]{30} \approx$ _____

# Distance on the Coordinate Plane

## Real-World Link

**Mountain Biking** Evan was biking on a trail. A map of the trail is shown. His brother timed his ride from point *A* to point *B*.

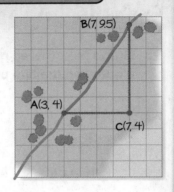

B(7, 9.5)

A(3, 4)

C(7, 4)

1. What do the blue and red lines on the graph represent?

_____

_____

_____

2. What type of triangle is formed by the lines?

_____

3. How can you find the length of $\overline{AC}$ and $\overline{BC}$ without counting the number of units?

_____

_____

4. What are the lengths of the two blue lines?

   $AC =$ ☐ units        $BC =$ ☐ units

5. Write an equation using the Pythagorean Theorem that you can use to find the length of $\overline{AB}$.

_____

Race ya!

**Essential Question**

HOW can algebraic concepts be applied to geometry?

**Vocabulary**

Distance Formula

**Common Core State Standards**

Content Standards
8.G.8, 8.EE.2

**MP Mathematical Practices**
1, 3, 4, 5

## Which MP Mathematical Practices did you use?
### Shade the circle(s) that applies.

① Persevere with Problems
② Reason Abstractly
③ Construct an Argument
④ Model with Mathematics
⑤ Use Math Tools
⑥ Attend to Precision
⑦ Make Use of Structure
⑧ Use Repeated Reasoning

# Find Distance on the Coordinate Plane

You can use the Pythagorean Theorem to find the distance between two points on the coordinate plane.

## Example

**1.** Graph the ordered pairs (3, 0) and (7, −5). Then find the distance c between the two points. Round to the nearest tenth.

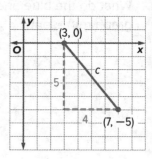

| | |
|---|---|
| $a^2 + b^2 = c^2$ | Pythagorean Theorem |
| $4^2 + 5^2 = c^2$ | Replace $a$ with 4 and $b$ with 5. |
| $41 = c^2$ | $4^2 + 5^2 = 16 + 25$ or 41 |
| $\pm\sqrt{41} = \sqrt{c^2}$ | Definition of square root |
| $\pm 6.4 \approx c$ | Use a calculator. |

The points are about 6.4 units apart.

**Distance**

To find the distance between two points on the coordinate plane, graph the points then draw a right triangle with c as the hypotenuse.

*Show your work.*

## Got It? Do this problem to find out.

a. _____

**a.** (1, 3), (−2, 4)

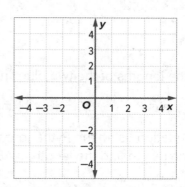

## Key Concept ▷ Distance Formula

**Symbols** The distance $d$ between two points with coordinates $(x_1, y_1)$ and $(x_2, y_2)$ is given by the formula
$$d = \sqrt{(x_2 - x_1)^2 + (y_2 - y_1)^2}.$$

**Model**

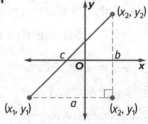

You can also use the **Distance Formula** to find the distance between two points on the coordinate plane. You can use the model from the Key Concept box to see how the Distance Formula is based on the Pythagorean Theorem as shown below.

$c^2 = a^2 + b^2$       Pythagorean Theorem

$c^2 = (x_2 - x_1)^2 + (y_2 - y_1)^2$       Substitute. The length of side $a$ is $(x_2 - x_1)$, and the length of side $b$ is $(y_2 - y_1)$.

$c = \sqrt{(x_2 - x_1)^2 + (y_2 - y_1)^2}$       Definition of square root

## Example

 Tutor

**2.** On the map, each unit represents 45 miles. West Point, New York, is located at (1.5, 2) and Annapolis, Maryland, is located at (−1.5, −1.5). What is the approximate distance between West Point and Annapolis?

**Method 1**

**Use the Pythagorean Theorem**

Let $c$ represent the distance between West Point and Annapolis. Then $a = 3$ and $b = 3.5$.

$a^2 + b^2 = c^2$

$3^2 + 3.5^2 = c^2$

$21.25 = c^2$

$\pm\sqrt{21.25} = \sqrt{c^2}$

$\pm 4.6 \approx c$

**Method 2**

**Use the Distance Formula**

Let $(x_1, y_1) = (1.5, 2)$ and $(x_2, y_2) = (−1.5, −1.5)$.

$c = \sqrt{(x_2 - x_1)^2 + (y_2 - y_1)^2}$

$c = \sqrt{(−1.5 − 1.5)^2 + (−1.5 − 2)^2}$

$c = \sqrt{(−3)^2 + (−3.5)^2}$

$c = \sqrt{9 + 12.25}$

$c = \sqrt{21.25}$

$c \approx \pm 4.6$

Since each map unit equals 45 miles, the distance between the cities is 4.6 · 45 or about 207 miles.

---

**Got It?** Do this problem to find out.

**b.** Cromwell Field is located at (2.5, 3.5) and Dedeaux Field at (1.5, 4.5) on a map. If each map unit is 0.1 mile, about how far apart are the fields?

 Show your work.

b. _____

## Example

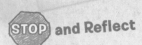
**3.** **Use the Distance Formula to find the distance between $X(5, -4)$ and $Y(-3, -2)$. Round to the nearest tenth if necessary.**

$$d = \sqrt{(x_2 - x_1)^2 + (y_2 - y_1)^2}$$   Distance Formula

$$XY = \sqrt{(-3 - 5)^2 + [-2 - (-4)]^2}$$   $(x_1, y_1) = (5, -4),$
$(x_2, y_2) = (-3, -2)$

$$XY = \sqrt{(-8)^2 + 2^2}$$   Simplify.

$$XY = \sqrt{64 + 4}$$   Evaluate $(-8)^2$ and $2^2$.

$$XY = \sqrt{68}$$   Add 64 and 4.

$$XY \approx \pm 8.2$$   Simplify.

So, the distance between points $X$ and $Y$ is about 8.2 units.

## Guided Practice

Check

**1.** Graph the ordered pairs (1, 5) and (3, 1). Then find the distance between the points. Round to the nearest tenth if necessary. (Example 1)

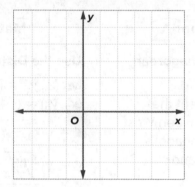

**2.** On a park map, the ranger station is located at (2.5, 3.5) and the nature center is located at (0.5, 4). Each unit in the map is equal to 0.5 mile. What is the approximate distance between the ranger station and the

nature center? (Examples 2 and 3) _____

**3.** @ **Building on the Essential Question** How can you use the Pythagorean Theorem to find the distance between two points on the coordinate plane?

_____

_____

_____

_____

**Rate Yourself!**

Are you ready to move on? Shade the section that applies.

YES   ?   NO

For more help, go online to access a Personal Tutor.   Tutor

FOLDABLES Time to update your Foldable!

# Independent Practice

Go online for Step-by-Step Solutions

**Graph each pair of ordered pairs. Then find the distance between the points. Round to the nearest tenth if necessary.** (Example 1)

**1.** (4, 5), (2, 2) ___3.6___

**2.** (−3, 4), (1, 3) ___4.1___

**3** (2.5, −1), (−3.5, −5)

___6.4___

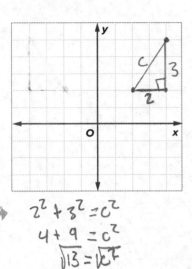

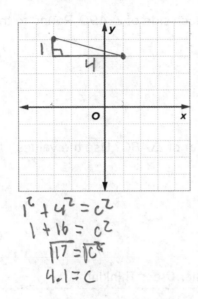

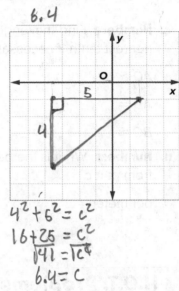

$2^2 + 3^2 = c^2$
$4 + 9 = c^2$
$\sqrt{13} = \sqrt{c^2}$
$3.6 = c$

$1^2 + 4^2 = c^2$
$1 + 16 = c^2$
$\sqrt{17} = \sqrt{c^2}$
$4.1 = c$

$4^2 + 6^2 = c^2$
$16 + 25 = c^2$
$\sqrt{41} = \sqrt{c^2}$
$6.4 = c$

**4.** A ferry sets sail from an island located at (4, 12) on a map. Its destination is Ferry Landing *B* at (6, 2). How far will the ferry travel if each unit on the grid is 0.5 mile? (Example 2) ___5.1 miles___

**Use the Distance Formula to find the distance between each pair of points. Round to the nearest tenth if necessary.** (Example 3)

**5** C(−5, −3), D(−4, −2)

___1.4___

**6.** Y(3.5, 1), Z(−4, 2.5)

___7.6___

**7.** $K\left(8\frac{1}{2}, 12\right)$, $L\left(-6\frac{3}{4}, 7\frac{1}{2}\right)$

___15.9___

**8.** Chicago, Illinois, has a longitude of 88°W and a latitude of 42°N. Indianapolis, Indiana, is located at 86°W and 40°N. At this longitude/latitude, each degree is about 53 miles. Find the distance between Chicago and

Indianapolis. ___150___

9. **MP** **Multiple Representations** Points $A(-2, 1)$, $B(-2, 6)$, and $C(1, 3)$ are the vertices of a triangle.

a. **Graphs** Graph the points $A(-2, 1)$, $B(-2, 6)$ and $C(1, 3)$.

b. **Words** Explain how to find the length of segment $BC$.

_____

_____

c. **Numbers** Find the length of each side of $\triangle ABC$. Round to the nearest tenth if necessary.

$AC \approx$ _____

$AB =$ _____

$BC \approx$ _____

d. **Numbers** What is the perimeter of $\triangle ABC$? Use the values from part c.

perimeter = _____

## H.O.T. Problems Higher Order Thinking

10. **MP** **Use Math Tools** Layla needs to find the distance between the points $A(-2.4, 3.7)$ and $B(4.5, -1.4)$. Suggest a tool she could use to find the length. Then find the length. Explain your reasoning.

_____

_____

11. **MP** **Persevere with Problems** Apply what you have learned about distance on the coordinate plane to write the coordinates of two possible endpoints of a line segment that is neither horizontal nor vertical and has a length of 5 units.

_____

_____

12. **MP** **Reason Inductively** Compare the steps to find the distance between two points on the coordinate plane by first using the Pythagorean Theorem and then using the Distance Formula.

_____

_____

_____

# Extra Practice

**Graph each pair of ordered pairs. Then find the distance between the points.**
**Round to the nearest tenth if necessary.**

**13.** (6, 2), (1, 0)  _5.4 units_

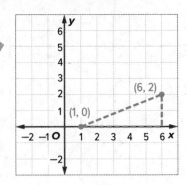

$a = 2, b = 5$

$a^2 + b^2 = c^2$

$2^2 + 5^2 = c^2$

$4 + 25 = c^2$

$29 = c^2$

$\pm\sqrt{29} = \sqrt{c^2}$

$\pm 5.4 \approx c$

**14.** (−5, 1), (2, 4)  _7.6_

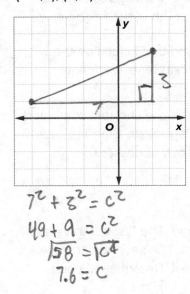

$7^2 + 3^2 = c^2$

$49 + 9 = c^2$

$\sqrt{58} = \sqrt{c^2}$

$7.6 = c$

**15.** (4, −2.3), (−1, −6.3)

_6.8_

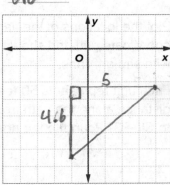

$4.6^2 + 5^2 = c^2$

$21.16 + 25 = c^2$

$\sqrt{46.16} = \sqrt{c^2}$

$6.8 = c$

**16.** **MP** **Use Math Tools** On a map of Minnesota, Mankato is located at
(3, 2.5), and Duluth is located at (8.5, 14.5). Each unit on the map equals
16.5 miles. What is the approximate distance between the cities?

_217.8_

**Use the Distance Formula to find the distance between each pair of points.**
**Round to the nearest tenth if necessary.**

**17.** W(1, 7), X(−2, −4)

_11.4_

**18.** G(−6.25, 5), H(−3.75, 2)

_3.9_

**19.** $P\left(-9\frac{1}{4}, -7\frac{1}{2}\right)$, Q(−4, 5)

_66.6_

**20.** Mr. Brady is using a coordinate plane to design a treasure hunt for his students. The hunt begins at the flagpole. The first clue is hidden 5 units north of the flagpole. The second clue is located 6 units east of the flagpole. Clue 2 says that Clue 3 is located 5 units south of the flagpole.

Plot the locations of the flagpole and the 3 clues on the coordinate grid and show the path students will follow with straight lines.

Each unit represents 50 feet. What is the shortest combined distance along the path from the flagpole to Clue 1 to Clue 2 to Clue 3? Round to the nearest foot if necessary.

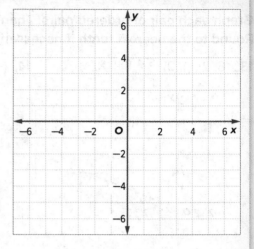

**21.** Two cars leave a house in Richmond. The first car travels 8 miles north and then 6 miles east. The second car travels 12 miles south and then 9 miles west. Determine if each statement is true or false.

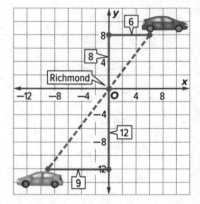

**a.** The first car is 10 miles from Richmond.    ☐ True ☐ False

**b.** The second car is 15 miles from Richmond.    ☐ True ☐ False

**c.** The cars are 35 miles apart.    ☐ True ☐ False

## CCSS Common Core Spiral Review

**Graph each set of points on the coordinate plane. Then connect the points and identify the figure drawn.** 6.G.3

**22.** $X(-1, -2)$, $Y(1, 2)$, $Z(3, -2)$

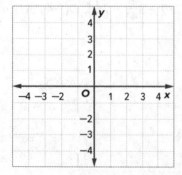

**23.** $A(0, 0)$, $B(-3, 4)$, $C(3, 4)$, $D(3, 0)$

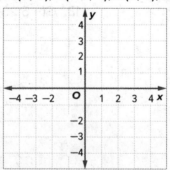

# 21ST CENTURY CAREER
## in Travel and Tourism

## Travel Agent

Do you love to travel? Would you be interested in helping others plan their ideal vacation getaways? You should consider a career in travel and tourism. Travel agents offer advice on destination locations and make arrangements for transportation, lodging, car rentals, and tours for their clients. In addition to having personal travel experience and being knowledgeable about popular vacation destinations, travel agents also need to be detail-oriented and have excellent communication, math, and computer skills.

## College & Career
### READINESS

Explore college and careers at ccr.mcgraw-hill.com

## Is This the Career for You?

Are you interested in a career as a travel agent? Take some of the following courses in high school.

- ◆ Algebra
- ◆ Business Software Applications
- ◆ Computer Technology
- ◆ Geometry

Turn the page to find out how math relates to a career in Travel and Tourism.

## MP Time to Get Away!

**Use the map to solve each problem. Round to the nearest tenth if necessary.**

1. What is the approximate distance between Key Largo and Islamorada? _____

2. Draw and label a right triangle to find the distance between Plantation Key and Islamorada. Then find the approximate distance. _____

3. Describe the ordered pairs that represent Layton and Plantation Key. Then find the approximate distance between Layton and Plantation Key. _____

4. To the nearest 0.5 unit, name the ordered pairs that represent Key West and Cudjoe Key. Then use the ordered pairs to estimate the distance between the keys.

   _____

   _____

5. What is the approximate distance between Key West and Layton? _____

6. What is the approximate distance between Tavernier and Big Pine Key? _____

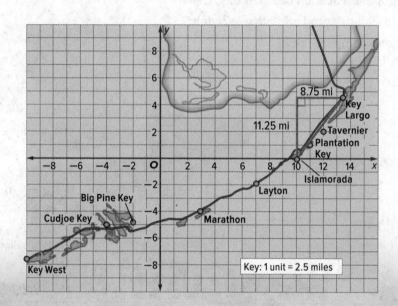

Key: 1 unit = 2.5 miles

## MP Career Project

It's time to update your career portfolio! Go to the Occupational Outlook Handbook online and research a career as a travel agent. Describe three things that you learned about being a travel agent that you did not know.

_____

_____

_____

_____

_____

List several jobs that are created by the travel and tourism industry.

- _____
- _____
- _____
- _____
- _____

# Chapter Review

## Vocabulary Check

Fill in the blank with the correct vocabulary term. Then circle the word that completes the sentence in the word seach.

1. A line that intersects two or more lines is called a _____.

2. _____ are those angles that are in the same position on the two lines in relation to the transversal.

3. _____ uses facts, rules, definitions, or laws to make conjectures from given situations.

4. A statement or conjecture that has been proven and can be used as a reason to justify statements in other proofs is called a _____.

5. The _____ describes the relationship between the lengths of the legs and the hypotenuse of *any* right triangle.

6. Interior angles that lie on opposite sides of the transversal are called _____.

7. Two lines that are in the same plane and do not intersect are called _____.

8. _____ is the process making a conjecture after observing several examples.

9. The _____ is the side opposite the right angle in a right triangle.

10. A polygon that is equilateral (all sides the same length) and equiangular (all angles have the same measure) is called a _____.

```
W E E F C A E U V M A T W W M Z L N W Y L H D H N
P Z Y L S J A B Z H P P R N Y P Z E L G N A I R O
W Q E K O Q Q Z W D K D J W H F E S M C Q F S O G
V G N I N O S A E R E V I T C U D N I X T E T G Y
S E L G N A R O I R E T N I E T A N R E T L A F L
S E N I L R A L U C I D N E P R E P C J S T N O O
Y T I T P Y T H A G O R E A N T H E O R E M C O P
C H Z Q S U O M K H N O T Y L O P R A L U G A R R
D E T K F B O Y J T F X Y U B Q C L X A G L F P A
S O S E L G N A G N I D N O P S E R R O C J O E L
G R H H Y P O T E N U S E N I L L E L L A R A P U
A E C Q A C X G B T E G G G N T G O V Q N E M W G
U M V Y A X L I M B X P Y B T L L V C W Y C U B E
O N K G N I N O S A E R E V I T C U D E D J L T R
R D Z D E M O T P S Q N Q A L A S R E V S N A R T
```

## Use Your FOLDABLES

**Use your Foldable to help review the chapter.**

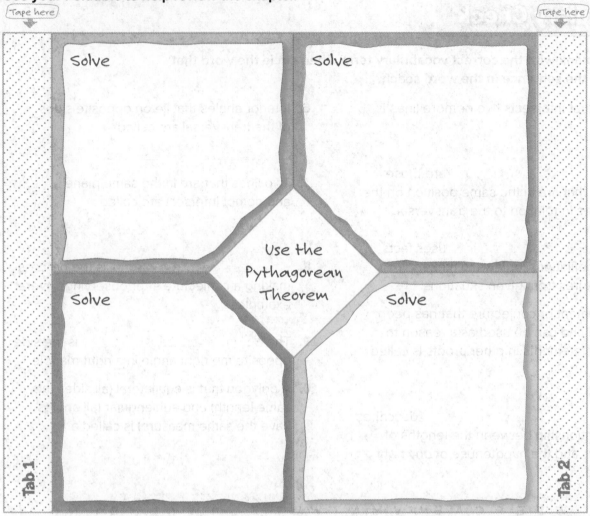

Tape here

Solve

Solve

Use the Pythagorean Theorem

Solve

Solve

Tab 1

Tab 2

Tape here

## Got it?

**Use the figure at the right. Circle the correct word to complete each sentence.**

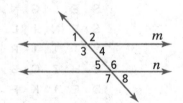

1. Angles 2 and 6 are examples of (vertical, **corresponding**) angles.

2. Angles 1 and 8 are examples of alternate (interior, **exterior**) angles.

3. The angle that is corresponding to angle 8 is angle (3, 4).

4. If lines *m* and *n* are parallel, then angles 3 and (6, 8) have equal measures.

# Power Up! Performance Task

## Under Construction

Midtown plans to build a highway through the city. The proposed highway will intersect Oakdale Road and Elm Street. Oakdale Road runs east to west, and Elm Street runs north to south so the two streets are perpendicular. The city created a proposal for a construction company as shown below.

| Proposal for Brook Highway |
| --- |
| • The distance on Elm Street from Oakdale Road to the highway will be three miles. |
| • The proposed highway will make a 32-degree angle with Oakdale Road. |
| • The distance on Oakdale Road from Elm Street to the highway will be 4.8 miles. |

**Write your answers on another piece of paper. Show all your work to receive full credit.**

### Part A
Draw a map that represents the proposed plan. Label the given information on the map. What is the measure of the angle that is between Elm Street and the proposed Brook Highway? Explain your answer.

### Part B
What will be the length of the section of Brook Highway from Oakdale Road to Elm Street? Round your answer to the nearest tenth of a mile.

### Part C
Mr. Shaw currently drives the Elm Street and Oakdale Road route five days a week, to and from work. If he gets 25 miles to a gallon of gasoline, how many gallons will he save every week when the highway is completed? Round to the nearest tenth.

### Part D
Walnut Street runs parallel to Oakdale Road and also intersects Brook Highway. The city requires the construction company to pay for a traffic signal if the intersection forms an angle of more 150°. Does the construction company pay for the signal? Explain your reasoning.

# Reflect

## Answering the Essential Question

Use what you learned about triangles and the Pythagorean Theorem to complete the graphic organizer. List three ways you used algebra in this chapter. Draw a model to represent each way.

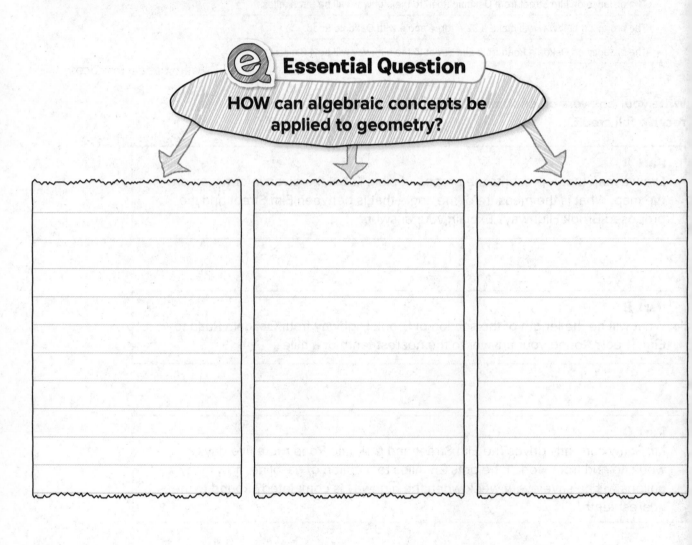

**Essential Question**

HOW can algebraic concepts be applied to geometry?

**Answer the Essential Question.** HOW can algebraic concepts be applied to geometry?

# Chapter 6
# Transformations

## Essential Question

HOW can we best show or describe the change in position of a figure?

## Common Core State Standards

**Content Standards**
8.G.1, 8.G.1a, 8.G.1b, 8.G.1c, 8.G.3

**MP Mathematical Practices**
1, 2, 3, 4, 5, 7, 8

## Math in the Real World

**Nature** Line symmetry occurs frequently in nature. A figure has line symmetry when a line can be drawn so that one half of the figure is a mirror image of the other other half.

On the figure below, draw the line of symmetry.

 **FOLDABLES®**
**Study Organizer**

**1** Cut out the Foldable on page FL5 of this book.

**2** Place your Foldable on page 498.

**3** Use the Foldable throughout this chapter to help you learn about transformations.

## Vocabulary

| | | |
|---|---|---|
| angle of rotation | dilation | reflection |
| center of dilation | image | rotation |
| center of rotation | line of reflection | rotational symmetry |
| congruent | preimage | transformation |
| | | translation |

## Review Vocabulary

**The Coordinate Plane** The x- and y-axes divide the coordinate plane into four regions called quadrants. Label the axes and the quadrants on the coordinate plane shown.

**Quadrilateral ABCD has vertices A(1, 1), B(3, 5), C(4, 7), and D(2, 6).**

1. In what quadrant is *ABCD* located? _____

2. Suppose you multiplied the coordinates of *ABCD* by $\frac{3}{4}$. In what quadrant would the

   new figure be located? _____

3. Suppose the *x*-coordinates in *ABCD* are multiplied by −1. In what quadrant would

   the new figure be located? _____

4. Suppose you switched the *x*- and *y*-coordinates from Exercise 3. In what quadrant

   would the new figure be located? _____

# What Do You Already Know?

Read each statement. Decide whether you agree (A) or disagree (D). Place a checkmark in the appropriate column and then justify your reasoning.

| Transformations | | | |
| --- | --- | --- | --- |
| Statement | A | D | Why? |
| When translating a figure, every point is moved the same distance. | | | |
| All figures have at least one line of symmetry. | | | |
| A reflection is a flip of a figure over a line of reflection. | | | |
| Rotations change a figure's orientation. | | | |
| Figures that have undergone a dilation are not congruent. | | | |
| When a figure is dilated by a scale factor less than 1, the dilated figure is larger than the original figure. | | | |

# When Will You Use This?

Here are a few examples of how transformations are used in the real world.

**Activity 1** How do you learn the newest dance crazes? Describe a dance that you like to do. How did you learn it? Was it easy to learn?

Alma and Carmen in
Dance Steps

**Activity 2** Go online at **connectED.mcgraw-hill.com** to read the graphic novel *Dance Steps*. How many steps of the dance are shown?

Try the Quick Check below.
Or, take the Online Readiness Quiz.

## Quick Review
**CCSS** **Common Core Review** 6.G.3, 7.NS.1

## Example 1

Two vertices of a rectangle are *J*(3, 2) and *K*(1, 2). The length of the rectangle is 4 units. Graph the rectangle and label the other two vertices.

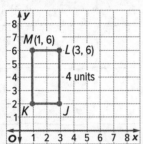

## Example 2

Find 2 + (−6).

2 + (−6) = −4        |2| − |−6| = −4
The sum is negative because |−6| > |2|.

## Quick Check

**Coordinate Plane**  Graph each figure and label the missing vertices.

1. rectangle with vertices:
   *B*(−3, 3), *C*(−3, 0); side length: 6 units

2. triangle with vertices:
   *Q*(−2, −4), *R*(2, −4); height: 4 units

3. square with vertices:
   *G*(5, 0), *H*(0, 5); side lengths: 5 units

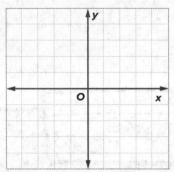

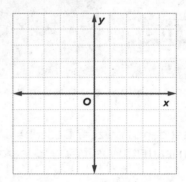

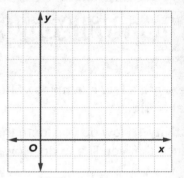

**Integers**  Add.

4. −5 + 3 = _____

5. 7 + (−9) = _____

6. −4 + (−9) = _____

7. −2 + 8 = _____

8. −8 + (−6) = _____

9. 0 + (−6) = _____

10. −8 + 2 = _____

11. 3 + (−1) = _____

## How Did You Do?

Which problems did you answer correctly in the Quick Check? Shade those exercise numbers below.

  ③ ④    ⑧  ⑩ ⑪

# Inquiry Lab
## Transformations

 **Inquiry** WHAT are some rigid motions of the plane?

 **Content Standards**
8.G.1, 8.G.1a, 8.G.1b, 8.G.1c

 **Mathematical Practices**
1, 3

Animated movies are created using frames. Each frame changes slightly from the previous one to create the impression of movement.

## Hands-On Activity 1

In this Activity, you will make animation frames using index cards.

**Step 1** Arrange ten index cards in a pile. On the top card, draw a circle at the top right hand corner.

**Step 2** On the next card, draw the same circle slightly down and to the left.

**Step 3** Repeat this for three or four more cards until your circle is at the bottom of the card. Use the remainder of the cards to draw the circle up and to the left.

**Step 4** Place a rubber band around the stack, hold the stack at the rubber band, and flip the cards from front to back.

Describe what you see when you flip the cards from front to back.

_____

_____

Look at the circles on the first and second cards and then the second and third cards. How would you describe the change in the position of the circle from one card to the next?

_____

_____

Did the shape or size of the circle change when you

moved it? If yes, describe the change. _____

# Hands-On Activity 2

**Step 1** Draw right angle *XYZ* on a piece of tracing paper. Place a dashed line on the paper as shown.

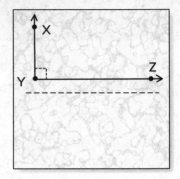

**Step 2** Fold the paper along the dashed line. Trace the angle onto the folded portion of the paper. Unfold and label the angle *ABC* so that *A* matches up with *X*, *B* matches up with *Y*, and *C* matches up with *Z*. Tape the paper to your book.

Use a protractor to find the measure of ∠*XYZ* and ∠*ABC*. Did the

measure of the angle change after the flip? _____

Use a centimeter ruler to measure the shortest distance from *X* and *A* to the dashed line. Repeat for *Y* and *B* and for *Z* and *C*. What do you notice?

_____

_____

# Hands-On Activity 3

**Step 1** Place a piece of tracing paper over the trapezoid shown. Copy the trapezoid. Draw points *A*, *B*, and *C*. Draw $\overrightarrow{AB}$.

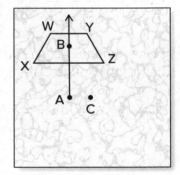

**Step 2** Place the eraser end of your pencil on *A*. Turn the tracing paper until $\overrightarrow{AB}$ passes through *C*. Tape the paper to your book.

Did the shape of the trapezoid change when you moved it? If yes, describe

the change. _____

Did the size of the trapezoid change when you moved

it? If yes, describe the change. _____

## Investigate

Collaborate

**Work with a partner. Use a ruler to draw the image when each figure is moved as directed.**

**1.** $\frac{1}{2}$ inch down and 1 inch to the left.

**2.** 1 inch up and 1 inch to the right.

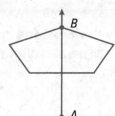

**Draw the image when each figure is flipped over line ℓ.**

**3.**

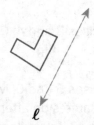

ℓ

**4.**

ℓ

**Draw the image when each pentagon is turned until $\overrightarrow{AB}$ passes through C.**

**5.**

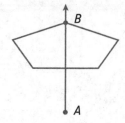

B

A

• C

**6.**

B

A

• C

**7.** Refer to Exercises 1–6.

   **a.** For which exercises, if any, did the size of the original figure change?

   _____

   **b.** For which exercises, if any, did the shape of the original figure change?

   _____

   **c.** For which exercises, if any, did the orientation of the original figure change?

   _____

# Analyze and Reflect

Collaborate

**For each pair of figures, describe a movement or movements that will place the blue figure on top of the green figure.**

8.

| Figure | Movement(s) |
| --- | --- |
|  |  |

9.

| Figure | Movement(s) |
| --- | --- |
|  |  |

10. Refer to Activity 1 and Exercises 1 and 2. Circle the word that best describes the movement of the figures:     **flip**          **slide**          **turn**

11. Refer to Activity 2 and Exercises 3 and 4. Circle the word that best describes the movement of the figures:     **flip**          **slide**          **turn**

12. Refer to Activity 3 and Exercises 5 and 6. Measure one side of the original figures. Then measure that same side after the turn. Did the length of the side change after you turned it? If yes, describe the change.

_____

13. **MP** **Justify Conclusions**  In Activity 3, $\overline{WY}$ and $\overline{XZ}$ are parallel. Were the segments still parallel after the turn? Would they still be parallel after a slide? flip? Explain.

_____

_____

_____

# Create

On Your Own

14. **MP** **Reason Inductively**  Slides, flips, and turns are called *rigid motions of the plane*. Based on the Activities, describe two characteristics of a rigid

motion of the plane. _____

_____

_____

15. **Inquiry**  WHAT are some rigid motions of the plane? _____

_____

# Lesson 1
# Translations

## Vocabulary Start-Up

A **transformation** is an operation that maps an original geometric figure, the **preimage**, onto a new figure called the **image**. A **translation** slides a figure from one position to another without turning it.

**Scan the lesson and complete the graphic organizer.**

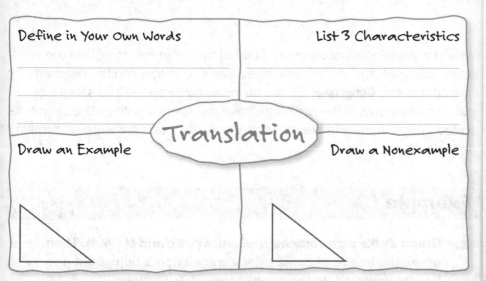

Define in Your Own Words

List 3 Characteristics

Translation

Draw an Example

Draw a Nonexample

## Real-World Link

Carmen created the design at the right on her computer.

1. Describe the motion involved in moving the design from $A$ to $A'$.

2. Compare the size, shape, and orientation of the design piece in its original position to that of the piece in its new position.

### Which MP **Mathematical Practices** did you use?
### Shade the circle(s) that applies.

① Persevere with Problems
② Reason Abstractly
③ Construct an Argument
④ Model with Mathematics

⑤ Use Math Tools
⑥ Attend to Precision
⑦ Make Use of Structure
⑧ Use Repeated Reasoning

# Translations in the Coordinate Plane

**Words** — When a figure is translated, the *x*-coordinate of the preimage changes by the value of the horizontal translation *a*. The *y*-coordinate of the preimage changes by the vertical translation *b*.

**Model**

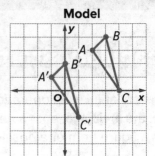

**Symbols** — $(x, y) \longrightarrow (x + a, y + b)$

**Prime Symbols**

Use prime symbols for vertices in a transformed image.

$A \rightarrow A'$

$B \rightarrow B'$

$B \rightarrow B'$

$A'$ is read *A prime.*

When translating a figure, every point of the preimage is moved the same distance and in the same direction. The image and the preimage are congruent. **Congruent** figures have the same shape and same size. So, line segments in the preimage have the same length as line segments in the image. Angles in the preimage have the same measure as angles in the image.

## Example

**1.** Graph △*JKL* with vertices *J*(−3, 4), *K*(1, 3), and *L*(−4, 1). Then graph the image of △*JKL* after a translation 2 units right and 5 units down. Write the coordinates of its vertices.

Move each vertex of the triangle 2 units right and 5 units down. Use prime symbols for the vertices of the image.

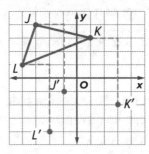  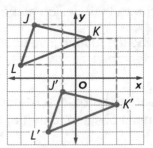

From the graph, the coordinates of the vertices of the image are *J*′(−1, −1), *K*′(3, −2), and *L*′(−2, −4).

**Got it?** Do this problem to find out.

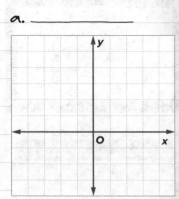

**a.** Graph △*ABC* with vertices *A*(4, −3), *B*(0, 2) and *C*(5, 1). Then graph its image after a translation of 4 units left and 3 units up. Write the coordinates of the image.

**a.** _____

## Example

Tutor

**2.** Triangle *XYZ* has vertices *X*(−1, −2), *Y*(6, −3) and *Z*(2, −5). Find the vertices of △*X'Y'Z'* after a translation of 2 units left and 1 unit up.

Use a table. Add −2 to the *x*-coordinates and 1 to the *y*-coordinates.

| Vertices of △*XYZ* | (*x* + (−2), *y* + 1) | Vertices of △*X'Y'Z'* |
|---|---|---|
| *X*(−1, −2) | (−1 + (−2), −2 + 1) | *X'*(−3, −1) |
| *Y*(6, −3) | (6 + (−2), −3 + 1) | *Y'*(4, −2) |
| *Z*(2, −5) | (2 + (−2), −5 + 1) | *Z'*(0, −4) |

So, the vertices of △*X'Y'Z'* are *X'*(−3, −1), *Y'*(4, −2), and *Z'*(0, −4).

## Got it? Do this problem to find out.

**b.** Quadrilateral *ABCD* has vertices *A*(0, 0), *B*(2, 0), *C*(3, 4), and *D*(0, 4). Find the vertices of quadrilateral *A'B'C'D'* after a translation of 4 units right and 2 units down.

Show your work.

b. _____

## Example

Tutor

**3.** A computer image is being translated to create the illusion of movement. Use translation notation to describe the translation from point *A* to point *B*.

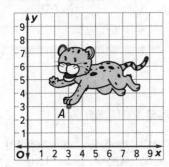

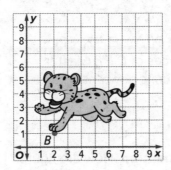

Point *A* is located at (3, 3). Point *B* is located at (2, 1).

$$(x, y) \rightarrow (x + a, y + b)$$

$$(3, 3) \rightarrow (3 + a, 3 + b) \rightarrow (2, 1)$$

$$3 + a = 2 \qquad\qquad 3 + b = 1$$

$$a = -1 \qquad\qquad b = -2$$

So, the translation is (*x* − 1, *y* − 2), 1 unit to the left and 2 units down.

**Got it?** Do this problem to find out.

c. Refer to the figure in Example 3. If point *A* was at (1, 5), use translation notation to describe the translation from point *A* to point *B*.

c. _____

## Guided Practice

Graph △*XYZ* with vertices *X*(−4, −4), *Y*(−3, −1), and *Z*(2, −2). Then graph the image of △*XYZ* after each translation, and write the coordinates of its vertices. (Example 1)

**1.** 3 units right and 4 units up

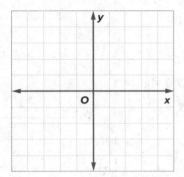

**2.** 2 units left and 3 units down

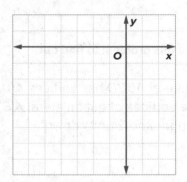

**3.** The baseball at the right was filmed using stop-motion animation so it appears to be thrown in the air. Use translation notation to describe the translation from point *A* to point *B*. (Example 3)

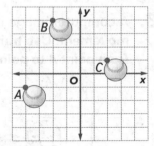

_____

**4.** Quadrilateral *DEFG* has vertices at *D*(1, 0), *E*(−2, −2), *F*(2, 4), and *G*(6, −3). Find the vertices of *D'E'F'G'* after a translation of 4 units right and 5 units down. (Example 2)

_____

**5.** ⓔ **Building on the Essential Question** How are figures translated on the coordinate plane?

_____

_____

_____

### Rate Yourself!

Are you ready to move on?
Shade the section that applies.

YES   ?   NO

For more help, go online to access a Personal Tutor.

Tutor 💬

FOLDABLES  Time to update your Foldable!

# Independent Practice

Go online for Step-by-Step Solutions

**Graph each figure with the given vertices. Then graph the image of the figure after the indicated translation, and write the coordinates of its vertices.** (Example 1)

**1** △*ABC* with vertices *A*(1, 2), *B*(3, 1), and *C*(3, 4) translated 2 units left and 1 unit up

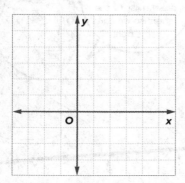

**2.** rectangle *JKLM* with vertices *J*(−3, 2), *K*(3, 5), *L*(4, 3), and *M*(−2, 0) translated 1 unit right and 4 units down

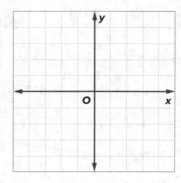

**Triangle *PQR* has vertices *P*(0, 0), *Q*(5, −2), and *R*(−3, 6). Find the vertices of *P′Q′R′* after each translation.** (Example 2)

**3.** 6 units right and 5 units up _____

**4.** 8 units left and 1 unit down _____

**Use the image of the race car at the right.** (Example 3)

**5.** Use translation notation to describe the translation from

point *A* to point *B*. _____

**6.** Use translation notation to describe the translation from

point *B* to point *C*. _____

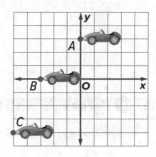

**7** Quadrilateral *KLMN* has vertices *K*(−2, −2), *L*(1, 1), *M*(0, 4), and *N*(−3, 5). It is first translated by $(x + 2, y − 1)$ and then translated by $(x − 3, y + 4)$. When a figure is translated twice, a double prime symbol is used. Find the coordinates of quadrilateral *K″L″M″N″* after both translations.

_____

8. **MP Model with Mathematics** Refer to the graphic novel frame below. List the five steps the girls should take and identify any transformations used in the dance steps. _____

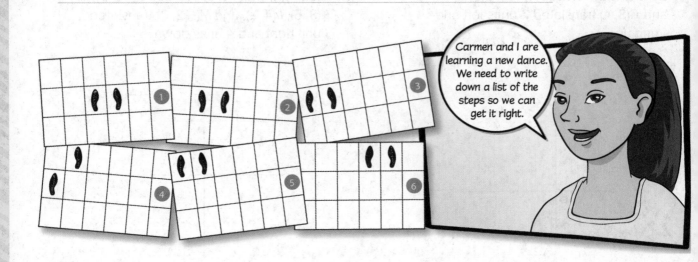

Carmen and I are learning a new dance. We need to write down a list of the steps so we can get it right.

## H.O.T. Problems Higher Order Thinking

9. **MP Reason Inductively** A figure is translated by $(x - 5, y + 7)$, then by $(x + 5, y - 7)$. Without graphing, what is the final position of the figure? Explain your reasoning to a classmate. _____

10. **MP Persevere with Problems** What are the coordinates of the point $(x, y)$ after being translated $m$ units left and $n$ units up? _____

11. **MP Reason Inductively** Determine whether each of the following statements is *always*, *sometimes*, or *never* true. Justify your reasoning.

   a. A translation preserves orientation. _____

   b. A preimage and its translated image are the same size, but not the same shape. _____

# Extra Practice

**Graph each figure with the given vertices. Then graph the image of the figure after the indicated translation, and write the coordinates of its vertices.**

**12.** △HJK with vertices H(−1, 0), J(−2, −4) and K(1, −3) translated 3 units right and 3 units up

H′(2, 3), J′(1, −1), K′(4, 0)

Graph each point and connect them to form a triangle. Then move each point 3 units to the right and then 3 units up. Connect them to form △H′J′K′.

**13.** Rectangle KLMN with vertices K(1, −1), L(1, 1), M(5, 1), and N(5, −1) translated 4 units left and 3 units up

_____

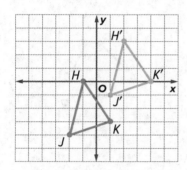

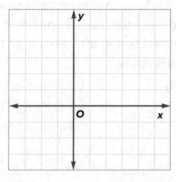

**Quadrilateral ABCD has vertices A(−5, −1), B(−3, 0), C(2, −2), and D(0, −6). Find the vertices of A′B′C′D′ after each translation.**

**14.** 4 units up _____

**15.** 2 units right and 2 units down _____

**16.** Julio is in Colorado exploring part of the Denver Zoo as shown. He starts at the Felines exhibit and travels 3 units to the right and 5 units up. At which exhibit is Julio located? If the Felines exhibit is located at (3, 1), what are the coordinates of Julio's new location?

_____

**17.** **MP** **Identify Repeated Reasoning** A diagram of a DNA double helix is shown below. Look for a pattern. On the diagram indicate where this pattern repeats or is translated. Find how many translations of the original pattern are shown in the diagram. _____

**18.** Graph triangle *DEF* with vertices *D*(−5, 2), *E*(1, 3), and
*F*(−4, −3). Then graph the image of the triangle after it
is translated 4 units right and 3 units down.

What are the vertices of triangle *D'E'F'*?

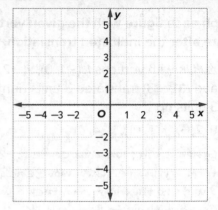

**19.** Trapezoid *ABCD* is shown on the coordinate plan. Suppose
the trapezoid is translated 3 units right and 2 units up.
Which of the following are vertices of the translated figure?
Select all that apply.

- ☐ *A'*(−2, 4)
- ☐ *B'*(1, −1)
- ☐ *C'*(0, 7)
- ☐ *D'*(0, 6)

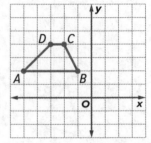

## (CCSS) **Common Core Spiral Review**

**Find each sum.** 7.NS.1

**20.** −5 + 12 =

**21.** 23 + (−3) = _____

**22.** −36 + (−42) = _____

**23.** 256 + (−82) = _____

**24.** −121 + (−119) = _____

**25.** −452 + 97 = _____

**Real-World Link**

**Art** Pysanky is the ancient Ukrainian art of egg decorating. Many artists use flips and line symmetry to create their designs. Use the activity to create your own pysanky design.

Collaborate

**The template shown represents the front view of an egg. The template has been divided into four sections.**

**Step 1** To create your egg, draw a design in Quadrant II.

**Step 2** To complete Quadrant I, draw the mirror image over the *y*-axis.

**Step 3** Repeat Steps 2 and 3 to fill in Quadrants III and IV. You can create a new design or you can draw the mirror image over the *x*-axis.

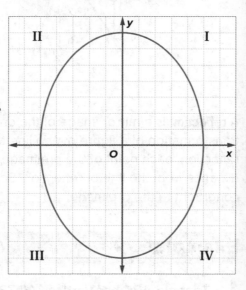

Add color to your design by using colored pencils or markers to complete the design.

1. *Line symmetry* is when a figure can be folded so one side is the mirror image of the other side. Does your pysanky have line symmetry? Explain. _____

_____

 **Essential Question**

HOW can you best show or describe the change in position of a figure?

 Vocab
**Vocabulary**

reflection
line of reflection

**Math Symbols**
$(x, y) \rightarrow (x, -y)$
$(x, y) \rightarrow (-x, y)$

CCSS **Common Core State Standards**

**Content Standards**
8.G.1, 8.G.3
MP **Mathematical Practices**
1, 3, 4, 7

**Which MP Mathematical Practices did you use?**
**Shade the circle(s) that applies.**

①  Persevere with Problems
②  Reason Abstractly
③  Construct an Argument
④  Model with Mathematics

⑤  Use Math Tools
⑥  Attend to Precision
⑦  Make Use of Structure
⑧  Use Repeated Reasoning

# Reflections in the Coordinate Plane

|  | Over the x-axis | Over the y-axis |
|---|---|---|
| **Words** | To reflect a figure over the x-axis, multiply the y-coordinates by −1. | To reflect a figure over the y-axis, multiply the x-coordinates by −1. |
| **Symbols** | $(x, y) \rightarrow (x, -y)$ | $(x, y) \rightarrow (-x, y)$ |
| **Models** | 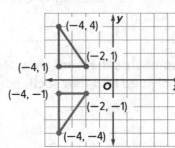 | 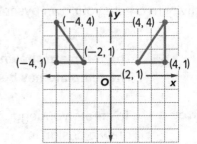 |

A **reflection** is a mirror image of the original figure. It is the result of a transformation of a figure over a line called a **line of reflection**. In a reflection, each point of the preimage and its image are the same distance from the line of reflection. So, in a reflection, the image is congruent to the preimage.

## Examples

**1.** Triangle *ABC* has vertices *A*(5, 2), *B*(1, 3), and *C*(−1, 1). Graph the figure and its reflected image over the x-axis. Then find the coordinates of the vertices of the reflected image.

The x-axis is the line of reflection. So, plot each vertex of *A′B′C′* the same distance from the x-axis as its corresponding vertex on *ABC*.

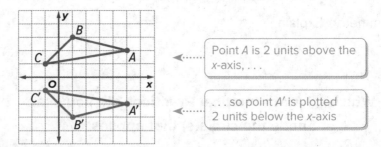

Point *A* is 2 units above the x-axis, . . .

. . . so point *A′* is plotted 2 units below the x-axis

The coordinates are *A′*(5, −2), *B′*(1, −3), and *C′*(−1, −1).

**Check**

Check the coordinates of the image by multiplying the y-coordinates by −1.

$(x, y) \rightarrow (x, -y)$

$(5, 2) \rightarrow (5, -2)$

$(1, 3) \rightarrow (1, -3)$

$(-1, 1) \rightarrow (-1, -1)$ ✔

**2.** Quadrilateral *KLMN* has vertices *K*(2, 3), *L*(5, 1), *M*(4, −2), and *N*(1, −1). Graph the figure and its reflection over the *y*-axis. Then find the coordinates of the vertices of the reflected image.

The *y*-axis is the line of reflection. So, plot each vertex of *K′L′M′N′* the same distance from the *y*-axis as its corresponding vertex on *KLMN*.

| Point *K′* is 2 units to the left of the *y*-axis. | 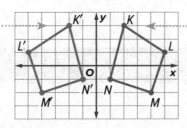 | Point *K* is 2 units to the right of the *y*-axis. |

The coordinates are *K′*(−2, 3), *L′*(−5, 1), *M′*(−4, −2), and *N′*(−1, −1).

---

**Got it?** Do this problem to find out.

**a.** Triangle *PQR* has vertices *P*(1, 5), *Q*(3, 7), and *R*(4, −1). Graph the figure and its reflection over the *y*-axis. Then find the coordinates of the reflected image.

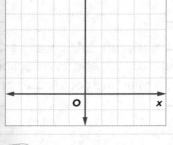

Show your work.

a. _____

---

## Example

Tutor

**3.** The figure below is reflected over the *y*-axis. Find the coordinates of point *A′* and point *B′*. Then sketch the figure and its image on the coordinate plane.

Point *A* is located at (1, 4). Point *B* is located at (2, 1). Since the figure is being reflected over the *y*-axis, multiply the *x*-coordinates by −1.

*A*(1, 4) → *A′*(−1, 4)          *B*(2, 1) → *B′*(−2, 1)

**STOP and Reflect**

Explain below how the x- and y-coordinates of an image relate to the x- and y-coordinates of the preimage after a reflection over the y-axis.

Show your work.

b. _____

Got it? Do this problem to find out.

b. The figure at the right is reflected over the *x*-axis. Find the coordinates of point *A′* and point *B′*. Then sketch the image on the coordinate plane.

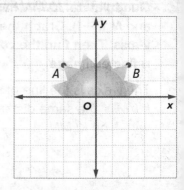

## Guided Practice

 Check ✓

1. Graph △*ABC* with vertices *A*(5, 1), *B*(1, 2), and *C*(6, 2) and its reflection over the *x*-axis. Then find the coordinates of the image.

(Examples 1 and 2)

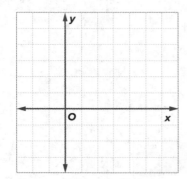

 Show your work.

2. The figure is reflected over the *y*-axis. Find the coordinates of point *A′* and point *B′*. Then sketch the image on the coordinate plane.

(Example 3)

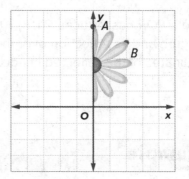

3.  **Building on the Essential Question** How can you determine the coordinates of a figure after a reflection over either axis?

_____

_____

_____

 **Rate Yourself!**

How well do you understand reflections? Circle the image that applies.

Clear    Somewhat Clear    Not So Clear

For more help, go online to access a Personal Tutor.   Tutor 💬

**FOLDABLES** Time to update your Foldable!

# Independent Practice

Go online for Step-by-Step Solutions

eHelp

**Graph each figure and its reflection over the indicated axis. Then find the coordinates of the reflected image.** (Examples 1 and 2)

**1** △GHJ with vertices G(4, 2), H(3, −4), and J(1, 1) over the y-axis

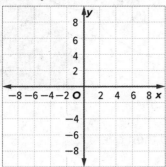

**2.** △MNP with vertices M(2, 1), N(−3, 1), and P(−1, 4) over the x-axis

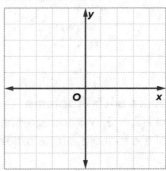

**3.** quadrilateral WXYZ with vertices W(−1, −1), X(4, 1), Y(4, 5), and Z(1, 7) over the x-axis

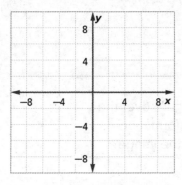

**4.** quadrilateral DEFG with vertices D(1, 0), E(1,−5), F(4, −1), and G(3, 2) over the y-axis

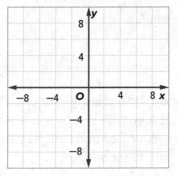

**5.** The figure at the right is reflected over the x-axis. Find the coordinates of point A′ and point B′. Then sketch the image on the coordinate plane.
(Example 3)

_____

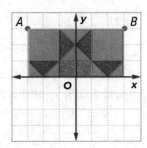

**MP** **Identify Structure** The coordinates of a point and its image after a reflection are given. Describe the reflection as over the x-axis or y-axis.

**6.** A(−3, 5) → A′(3, 5) _____

**7** M(3, 3) → M′(3, −3) _____

## H.O.T. Problems Higher Order Thinking

8. **MP** **Find the Error** Roberto is finding the coordinates of the image of a triangle with vertices A(1, 1), B(4, 1) and C(1, 5) after a reflection over the x-axis. Describe his mistake and correct it.

_____

_____

_____

The vertices of triangle A' B' C' are A'(−1, 1), B'(−4, 1) and C'(−1, 5).

9. **MP** **Persevere with Problems** Triangle JKL has vertices J(−7, 4), K(7, 1), and L(2, −2). Without graphing, find the new coordinates of the vertices of the triangle after a reflection first over the x-axis and then over the y-axis. _____

_____

_____

10. **MP** **Reason Inductively** Suppose you reflect a triangle in Quadrant I over the y-axis, then translate the image 2 units left and 3 units down. Is there a single transformation that maps the preimage onto the image? Explain your reasoning. _____

_____

_____

11. **MP** **Reason Inductively** Suppose you reflect a nonregular figure over the x-axis and then reflect it over the y-axis. Is there a single transformation using reflections or translations that maps the preimage onto the image? Explain your reasoning.

_____

_____

_____

_____

12. **MP** **Which One Doesn't Belong?** Triangle ABC has vertices A(1, 2), B(1, 5), and C(4, 2) and undergoes a transformation. Circle the set of vertices that does not belong. Explain your reasoning.

| A'(1, −1), B'(1, 2), C'(4, −1) | A'(5, 2), B'(5, 5), C'(8, 2) |
|---|---|

| A'(1, −2), B'(1, −5), C'(4, −2) | A'(3, 3), B'(3, 6), C'(6, 3) |
|---|---|

_____

_____

# Extra Practice

**Graph each figure and its reflection over the indicated axis. Then find the coordinates of the reflected image.**

**13.** △*TUV* with vertices *T*(−4, −1), *U*(−2, −3), and *V*(4, −3) over the *x*-axis

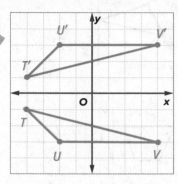

T′(−4, 1), U′(−2, 3), V′(4, 3)

**14.** square *ABCD* with vertices *A*(2, 4), *B*(−2, 4), *C*(−2, 8), and *D*(2, 8) over the *x*-axis

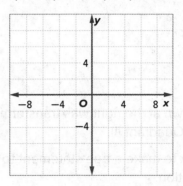

**15.** △*RST* with vertices *R*(−5, 3), *S*(−4, −2), and *T*(−2, 3) over the *y*-axis

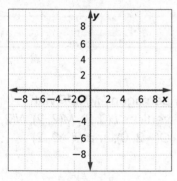

**16.** parallelogram *HIJK* with vertices *H*(−1, 3), *I*(−1, −1), *J*(2, −2), and *K*(2, 2) over the *y*-axis

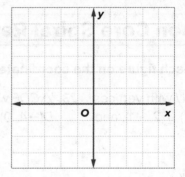

**17.** The figure at the right is reflected over the *y*-axis. Find the coordinates of point *A′* and point *B′*. Then sketch the image on the coordinate plane.

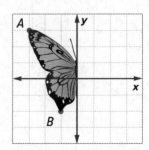

_____

**MP Identify Structure** The coordinates of a point and its image after a reflection are given. Describe the reflection as over the *x*-axis or *y*-axis.

**18.** *X*(−1, −4) → *X′*(−1, 4) _____

**19.** *W*(−4, 0) → *W′*(4, 0) _____

**20.** Graph the image of triangle *RST* after it is reflected over the *x*-axis then translated 4 units to the right and 3 units down.

What are the vertices of triangle *R'S'T'*?

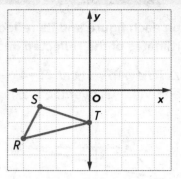

**21.** The figure shown at the right was transformed from Quadrant II to Quadrant III.

Fill in each box to make a true statement to describe the transformation.

The figure was [          ] over the [          ].

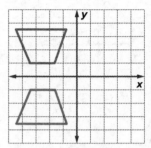

**CCSS** **Common Core Spiral Review**

**Graph and label each figure on the coordinate plane.** 6.G.3

**22.** trapezoid *RSTU* with vertices *R*(−1, 3), *S*(2, 2), *T*(2, −1), and *U*(−1, −2)

**23.** △*CDE* with vertices *C*(−1, 1), *D*(−4, 0), and *E*(0, −2)

**24.** pentagon *LMNOP* with vertices *L*(0, 3), *M*(2, 2), *N*(2, 0), *O*(−2, 0), and *P*(−2, 2)

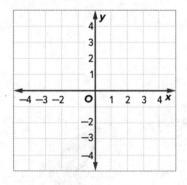

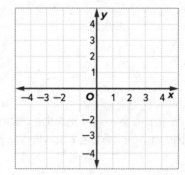

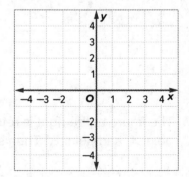

**Problem-Solving Investigation**
# Act It Out

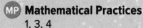 **Content Standards**
8.G.3

MP **Mathematical Practices**
1, 3, 4

## Case #1 Black Belt Champion

Emily's school is 3 blocks east and 4 blocks south from her house. She is taking a martial arts class that is 2 blocks east and 6 blocks north from school.

*What are two different ways that Emily can travel to go from martial arts class to her house?*

## Understand *What are the facts?*

You know the translations involved.
- School is 3 blocks east and 4 blocks south from her house.
- Martial arts class is 2 blocks east and 6 blocks north from school.

## Plan *What is your strategy to solve this problem?*

Act out the situation on a coordinate plane. Plot Emily's house at (0, 0) and map out the route to her school and martial arts class. Then determine two translations that will take Emily from martial arts class to her house.

## Solve *How can you apply the strategy?*

What are two different ways that Emily can travel to go from martial arts class to her house?

_____

_____

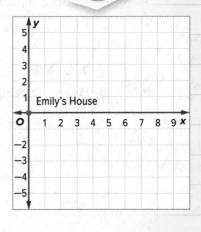

## Check *Does the answer make sense?*

Begin with the point (0, 0) to represent Emily's house. Use translation notation to determine the route to school, martial arts class and then back home.

## Analyze the Strategy  Tutor

MP **Make a Conjecture** Suppose Emily needed to drive 32 blocks east and 15 blocks north from school. Would it be more efficient using translation notation or acting the problem out on graph paper? Explain.

_____

_____

## Case #2  Keep the Change

Shiro bought an apple juice and a bag of pretzels for $4.55.

If he paid the cashier with a $5 bill, in how many different ways can he receive his change if the cashier only gives him quarters, dimes, and nickels?

# 1 Understand

**Read the problem. What are you being asked to find?**

I need to find _____.

**Underline key words and values. What information do you know?**

Shiro's purchase was ☐ and he paid with a ☐ bill. The change is in _____ , _____ , and _____ .

# 2 Plan

**Choose a problem-solving strategy.**

I will use the _____ strategy.

# 3 Solve

**Use your problem-solving strategy to solve the problem.**

Use counters or coins to represent _____ , _____ , and _____ . Because Shiro received ☐ in change, use the coins to find different combinations with a sum of ☐ . Record each combination. Q = quarters, D = dimes, and N = nickels.

Combinations possible: 1Q, ☐ D; 1Q, 1D, ☐ N; 1Q, 4N; ☐ D, 1 N; ☐ D, 3 N; 2 D, ☐ N; 1 D, 7 N; ☐ N.

So, _____.

# 4 Check

**Use information from the problem to check your answer.**

_____

Collaborate

**Work with a small group to solve the following cases.
Show your work on a separate piece of paper.**

## Case #3  Picture Exchange

The French Club took a field trip to an exhibit of French art at the museum. Five of the club members held a picture exchange to share their pictures. April brought more pictures than Brandon. Chloe brought more pictures than Diego, but fewer than Brandon. Ethan brought more pictures than Chloe, but not as many as Brandon.

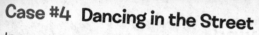

List the picture exchange participants in order from the most pictures to the fewest.

## Case #4  Dancing in the Street

In a certain dance for a competition, a dancer makes the following series of steps: 2 steps back, 1 step to the right, 3 steps forward, 2 steps to the right. The series is repeated four times.

How does the dancer's final position compare to his original position?

## Case #5  Fitness

The length of a basketball court is 84 feet. Kareem starts at one end of the court and runs 20 feet forward and then 8 feet back.

How many more times will he have to do this until he reaches the end of the basketball court? What equation represents this relationship?

Use any strategy!

## Case #6  Parties

Abby sent a text message to three friends inviting each of them to a party. Each of those friends sent the message to three more friends. Then each of these friends sent the message to three more friends.

If two thirds of the friends receiving the text message attended the party, how many friends attended the party?

## Vocabulary Check

1. **MP Be Precise** Define *transformation* using the words *preimage* and *image*. (Lesson 1)

_____

_____

2. Describe the role of the line of reflection in a transformation. (Lesson 2)

_____

_____

## Skills Check and Problem Solving

**Graph each triangle with the given vertices. Then graph the image after the given transformation and write the coordinates of the image's vertices.**
(Lessons 1 and 2)

3. △*ABC* with vertices *A*(3, 5), *B*(4,1), and *C*(1, 2); translation of 3 units left and 4 units down

_____

4. △*WXY* with vertices *W*(−1, −2), *X*(0, −4), and *Y*(−3, −5); reflection over the *x*-axis followed by a reflection over the *y*-axis

_____

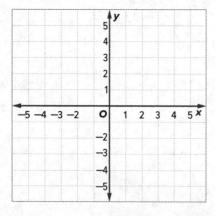

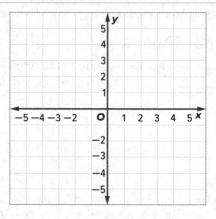

5. **MP Persevere with Problems** Point *D* is translated 5 units right and 2 units down, then reflected over the *y*-axis. Write an algebraic representation to represent the final location of point *D*. (Lessons 1 and 2)

_____

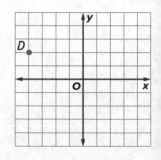

# Inquiry Lab
## Rotational Symmetry

CCSS Content Standards
8.G.1

MP Mathematical Practices
1, 3

**Inquiry** HOW can you identify rotational symmetry?

Many products have logos so people can easily identify them. If you turn the first aid logo below 180°, will the logo look the same as the original figure?

## Hands-On Activity

A figure has **rotational symmetry** if it can be rotated or turned less than 360° about its center so that the figure looks exactly as it does in its original position.

First aid box

**Step 1**  Copy the outline of the first aid logo onto a piece of tracing paper. Label one vertex *A*.

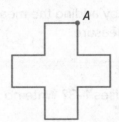

**Step 2**  Place the tracing paper over the outline in Step 1. Put your pencil point at the center of the figure to hold the tracing paper in place. Turn the tracing paper clockwise from its original position until the two figures match. Draw and label the new figure in the space provided.

**Step 3**  Continue turning the tracing paper until the logo is back to its original position. Does the figure have rotational symmetry? Explain.

_____

_____

**Investigate**

Work with a partner. Determine whether the figure has rotational symmetry. Write *yes* or *no*.

1.

2.

3.

 **Analyze and Reflect**

4. **MP Reason Inductively** The degree measure of an angle through which the figure is rotated is called the **angle of rotation**. Find the first angle of rotation of the first aid logo by dividing 360° by the total number of times the figures matched. _____

5. List the other angles of rotation of the first aid logo by adding the measure of the first angle of rotation to the previous angle measure.

   Stop when you reach 360°. _____

6. What is the angle of rotation of each figure in Exercises 1–3? Write *no* if there is no rotational symmetry.

   Exercise 1 _____    Exercise 2 _____    Exercise 3 _____

 **Create**

7. **MP Model with Mathematics** Draw two figures, one that has rotational symmetry and one that does not.

8. **Inquiry** HOW can you identify rotational symmetry?

   _____

   _____

   _____

# Rotations

 **Real-World Link**

**Prizes** Pablo is spinning the prize wheel shown below.

1. A spin can be *clockwise* or *counterclockwise*. Define these two words in your own words.

   clockwise _____

   counterclockwise _____

2. If the section labeled 8 on the left part of the wheel spins 90°

   clockwise, where will it land? _____

3. If one of the sections labeled 4 makes three complete turns counterclockwise, how many degrees will it have

   traveled? [        ]°

4. Are there any points on the wheel that stay fixed, or do not move, when the wheel spins? If so, what are the points?

   _____

5. Does the center of the wheel change if the wheel is spun

   counterclockwise as opposed to clockwise? _____

6. Does the distance from the center to the edge change as it spins? Explain.

   _____

   _____

 **Essential Question**

HOW can we best show or describe the change in position of a figure?

 **Vocabulary**

rotation
center of rotation

**Math Symbols**
$(x, y) \rightarrow (y, -x)$
$(x, y) \rightarrow (-x, -y)$
$(x, y) \rightarrow (-y, x)$

 **Common Core State Standards**

**Content Standards**
8.G.1, 8.G.3
**MP Mathematical Practices**
1, 3, 4, 7

**Which MP Mathematical Practices did you use?**
**Shade the circle(s) that applies.**

① Persevere with Problems
② Reason Abstractly
③ Construct an Argument
④ Model with Mathematics
⑤ Use Math Tools
⑥ Attend to Precision
⑦ Make Use of Structure
⑧ Use Repeated Reasoning

# Rotate a Figure About a Point

A **rotation** is a transformation in which a figure is rotated, or turned, about a fixed point. The **center of rotation** is the fixed point. A rotation does not change the size or shape of the figure. So, the preimage and the image are congruent.

## Rotations

Rotations can be described in degrees and direction. For example, 90° clockwise or 270° counterclockwise.

### Real World Example

**1.** Triangle *LMN* with vertices *L*(5, 4), *M*(5, 7), and *N*(8, 7) represents a desk in Jackson's bedroom. He wants to rotate the desk counterclockwise 180° about vertex *L*. Graph the figure and its image. Then give the coordinates of the vertices for △*L'M'N'*.

**Step 1**  Graph the original triangle.

**Step 2**  Graph the rotated image. Use a protractor to measure an angle of 180° with *M* as one point on the ray and *L* as the vertex. Mark off a point the same length as $\overline{ML}$. Label this point *M'* as shown.

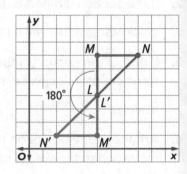

**Step 3**  Repeat Step 2 for point *N*. Since *L* is the point at which △*LMN* is rotated, *L'* will be in the same position as *L*.

So, the coordinates of the vertices of △*L'M'N'* are *L'*(5, 4), *M'*(5, 1), and *N'*(2, 1).

### Got it? Do this problem to find out.

Show your work.

a. Rectangle *ABCD* with vertices *A*(−7, 4), *B*(−7, 1), *C*(−2, 1), and *D*(−2, 4) represents the bed in Jackson's room. Graph the figure and its image after a clockwise rotation of 90° about vertex *C*. Then give the coordinates of the vertices for rectangle *A'B'C'D'*.

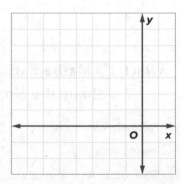

a. _____

# Rotations About the Origin

Key Concept

**Words**   A rotation is a transformation around a fixed point. Each point of the original figure and its image are the same distance from the center of rotation.

**Models**   The rotations shown are clockwise rotations about the origin.

| 90° Rotation | 180° Rotation | 270° Rotation |
|---|---|---|

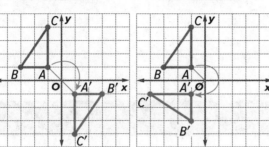

**Symbols**

$(x, y) \rightarrow (y, -x)$        $(x, y) \rightarrow (-x, -y)$        $(x, y) \rightarrow (-y, x)$

Figures can also be rotated about the origin.

# Example

Tutor

**2.** Triangle *DEF* has vertices *D*(−4, 4), *E*(−1, 2), and *F*(−3, 1). Graph the figure and its image after a clockwise rotation of 90° about the origin. Then give the coordinates of the vertices for △*D′E′F′*.

**Check**
Check the coordinates of the image.
$(x, y) \rightarrow (y, -x)$
$(-4, 4) \rightarrow (4, 4)$
$(-1, 2) \rightarrow (2, 1)$
$(-3, 1) \rightarrow (1, 3)$ ✔

**Step 1**   Graph △*DEF* on a coordinate plane.

**Step 2**   Sketch segment $\overline{EO}$ connecting point *E* to the origin. Sketch another segment, $\overline{E'O}$, so that the angle between point *E*, *O*, and *E′* measures 90° and the segment is the same length as $\overline{EO}$.

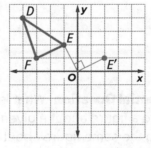

**Step 3**   Repeat Step 2 for points *D* and *F*. Then connect the vertices to form △*D′E′F′*.

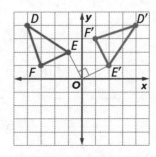

So, the coordinates of the vertices of △*D′E′F′* are *D′*(4, 4), *E′*(2, 1), and *F′*(1, 3).

**b.** _____

**b.** Quadrilateral *MNPQ* has vertices *M*(2, 5), *N*(6, 4), *P*(6, 1), and *Q*(2, 1). Graph the figure and its image after a counterclockwise rotation of 270° about the origin. Then give the coordinates of the vertices for quadrilateral *M'N'P'Q'*.

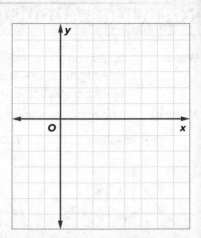

# Guided Practice

Triangle *XYZ* has vertices *X*(3, −1), *Y*(5, −4), and *Z*(1, −5). Graph △*XYZ* and its image after each rotation. Then give the coordinates of the vertices for △*X'Y'Z'*. (Examples 1 and 2)

**1.** 270° counterclockwise about vertex *X*

Show your work.

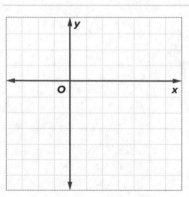

**2.** 180° clockwise about the origin

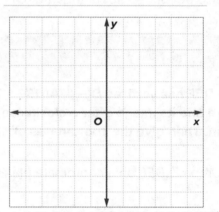

**3.** 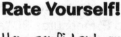 **Building on the Essential Question** What is the difference between rotating a figure about a given point that is a vertex and rotating the same figure about the origin if the rotation is less than 360°?

_____

_____

_____

_____

_____

_____

**Rate Yourself!**

How confident are you about rotations? Check the box that applies.

For more help, go online to access a Personal Tutor.

 Time to update your Foldable!

# Independent Practice

 Go online for Step-by-Step Solutions

**1.** Triangle *RST* represents the placement of Tyra's tricycle in the driveway and has vertices $R(-7, 8)$, $S(-7, 2)$, and $T(-2, 2)$. Graph the figure and its rotated image after a clockwise rotation of 180° about the origin. Then give the coordinates of the vertices for triangle *R'S'T'*. (Example 2)

_____

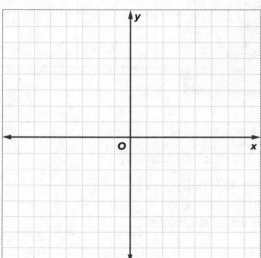

**2.** Quadrilateral *ABCD* has vertices at $A(-3, -4)$, $B(-1, -1)$, $C(2, -2)$, and $D(3, -4)$. Graph quadrilateral *ABCD* and its image after a 90° clockwise rotation about vertex *A*. Then give the coordinates of the vertices of the image. (Example 1)

_____

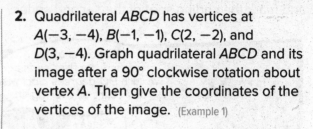

**3.**  **Model with Mathematics** A partial hubcap is shown. Copy and complete the figure so that the completed hubcap has rotational symmetry of 90°, 180°, and 270°.

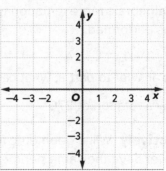

**4.** The right isosceles triangle *PQR* has vertices $P(3, 3)$, $Q(3, 1)$, and $R(x, y)$ and is rotated 90° counterclockwise about the origin. Find the missing vertex of the triangle. Then graph the triangle and its image.

$R(x, y) = R($_____, _____$)$

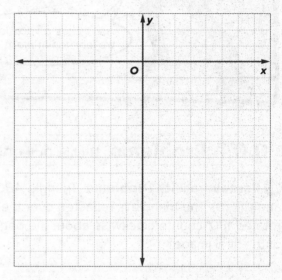

**5.** Which capital letters in VIRGINIA produce the same letter after being rotated 180°? _____

6. **MP Model with Mathematics** Refer to the graphic novel frame below. The last step is shown on grid 6. The girls make a clockwise rotation of 90° and begin the dance again. On a separate sheet of paper, expand the grid and mark the ending spot of the second series.

## H.O.T. Problems Higher Order Thinking

7. **MP Persevere with Problems** Triangle *ABC* has vertices *A*(0, 4), *B*(0, −2), and *C*(2, 0). The triangle is reflected over the *x*-axis. Then the image is rotated 180° counterclockwise about the origin. What are the coordinates of the final image?

8. **MP Persevere with Problems** Triangle *QRS* is translated 7 units right, then rotated 90° clockwise about the origin. The vertices of triangle *Q″R″S″* are *Q″*(6, −1), *R″*(0, −1), and *S″*(0, −7). Find the coordinates of △*QRS*.

9. **MP Model with Mathematics** A triangle is rotated 90° clockwise about the origin. Then the image is rotated 270° clockwise about the origin.

   a. Complete the algebraic representation to explain the effect of the series of transformations performed.

   $(x, y) \rightarrow ($ ____ , ____ $) \rightarrow ($ ____ , ____ $)$

   b. Based on your answer to part a, what can you conclude about a rotation of 90° followed by a rotation of 270°. _____

10. **MP Reason Inductively** Will a geometric figure and its rotated image *always*, *sometimes*, or *never* have the same perimeter? Explain your reasoning.

# Extra Practice

**11.** Quadrilateral *EFGH* has vertices *E*(1, −1), *F*(3, −5), *G*(7, −5), and *H*(6, −1). Graph the figure and its rotated image after a counterclockwise rotation of 90° about the origin. Then give the coordinates of the vertices for quadrilateral *E′F′G′H′*.

**12.** Quadrilateral *ABCD* has vertices at *A*(−3, −4), *B*(−1, −1), *C*(2, −2), and *D*(3, −4). Graph quadrilateral *ABCD* and its image after a 180° counterclockwise rotation about vertex *D*. Then give the coordinates of the vertices of the image.

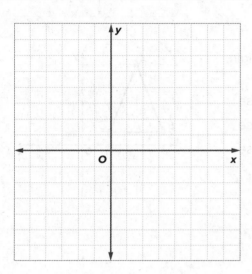

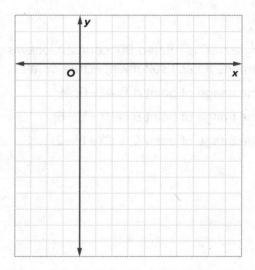

**13.** MP **Identify Structure** Identify each transformation as a *translation*, *reflection*, or *rotation*.

a.

b.

c.

**Copy and Solve** Triangle *MNP* has vertices *M*(1, 4), *N*(3, 1), and *P*(5, 3). Find the vertices of *M′N′P′* after each rotation about the origin. Show your work on a separate piece of paper.

**14.** 90° clockwise

**15.** 180° clockwise

**16.** 90° counterclockwise

**17.** On a floor plan, *TUVW* with vertices *T*(−4, 0), *U*(−4, 2), *V*(−1, 2), and *W*(−1, 0) represents the location of Chantal's bed in her bedroom. Chantal would like to rotate her bed 180° clockwise about point *V* to see if she likes the new placement. Draw the bed and the rotated image on the coordinate plane.

What are the coordinates of the corners of the rotated bed?

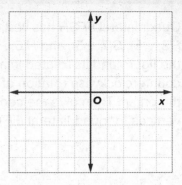

**18.** Triangle *ABC* is rotated 90° counterclockwise about the origin. Determine if each statement is true or false.

 **a.** The image of point *A* is *A*′(−6, 4).    ☐ True ☐ False

 **b.** The image of point *B* is *B*′(−1, −6).    ☐ True ☐ False

 **c.** The image of point *C* is *C*′(−1, −2).    ☐ True ☐ False

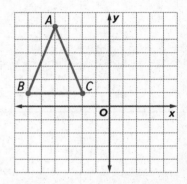

## CCSS Common Core Spiral Review

**19.** Use the graph of △*ABC* shown at the right. **8.G.1**
 **a.** What are the coordinates of △*A*′*B*′*C*′ when △*ABC* is reflected

 over the *x*-axis? _____

 **b.** Graph the image of △*ABC* after it is translated 2 units right and 1 unit up.

**20.** Triangle *FGH* has vertices *F*(−3, 7), *G*(−1, 5), and *H*(−2, 2). Find the vertices of its image after a translation of 4 units right and 2 units down followed by a reflection over the *y*-axis. **8.G.1**

_____

# Inquiry Lab
## Dilations

 **Inquiry** **WHAT are the results of a dilation of a triangle?**

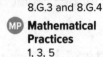 **Content Standards** Preparation for 8.G.3 and 8.G.4

**MP Mathematical Practices** 1, 3, 5

One way to create murals on a wall is to use a drawing grid method. Artists draw a grid on the artwork to be copied and draw a similar grid on the wall. By transferring sections of the artwork, the mural is the same shape as the artwork, but a different size.

## Hands-On Activity 1

In this Activity, you will enlarge △ABC by a *scale factor* of 2 using grid paper. Point A will be the center point for the enlargement.

**Step 1** On the grid shown below, $\overrightarrow{AB}$ is drawn to the edge of the grid. Draw $\overrightarrow{AC}$ in the same way.

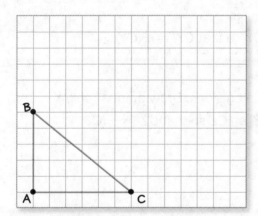

**Step 2** Draw point B′ on $\overrightarrow{AB}$ so that $AB' = 2(AB)$. Draw point C′ on $\overrightarrow{AC}$ so that $AC' = 2(AC)$.

**Step 3** Draw $\overline{B'C'}$ to complete △AB′C′.

What is the ratio of the length of $\overline{AB}$ to the length of $\overline{AB'}$? _____

What is the ratio of the length of $\overline{AC}$ to the length of $\overline{AC'}$? _____

What is the ratio of the length of $\overline{BC}$ to the length of $\overline{BC'}$? _____

What do you notice about the ratios of corresponding sides?

Is △ABC similar to △A′B′C′? _____

# Hands-On Activity 2

In Activity 1, you used a dilation to transform $\triangle ABC$ by a scale factor of 2. A **dilation** is a transformation that enlarges or reduces a figure by a scale factor relative to a center point. That point is called the **center of dilation**.

In this Activity, you will draw the image of $\triangle XYZ$ after a dilation with a scale factor of $\frac{1}{2}$. Point $C$ will be the center of dilation.

**Step 1**    Triangle $XYZ$ is shown below. Point $C$ is the center of dilation. Using a ruler, draw line segments connecting $C$ to each of the vertices of the triangle. $\overline{CY}$ is done for you.

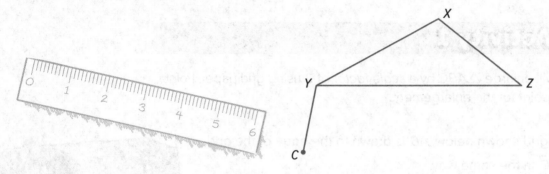

**Step 2**    Measure $\overline{CY}$. Draw point $Y'$ on $\overline{CY}$ so that $CY' = \frac{1}{2}(CY)$.

**Step 3**    Repeat Step 2 for the two remaining sides. Draw point $X'$ on $\overline{CX}$ so that $CX' = \frac{1}{2}(CX)$ and point $Z'$ on $\overline{CZ}$ so that $CZ' = \frac{1}{2}(CZ)$.

**Step 4**    Draw $\triangle X'Y'Z'$.

Is $\triangle X'Y'Z'$ the same shape as $\triangle XYZ$? _____

Measure and compare the corresponding lengths on the original and new triangles. Describe the relationship between these

measurements. _____

_____

Measure and compare the corresponding angles on the original and new triangles. Describe the relationship between these

measurements. _____

_____

# Investigate

Collaborate

**Work with a partner. Draw the image after a dilation with the given scale factor. Point _A_ is the center of dilation.**

**1.** scale factor: 3

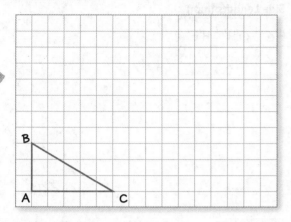

**2.** scale factor: $\frac{1}{3}$

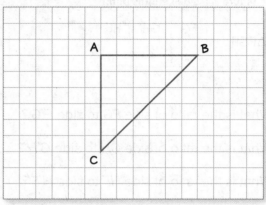

**Work with a partner. Use a ruler to draw the image after a dilation with the given scale factor. Point _C_ is the center of dilation.**

**3.** scale factor: 3

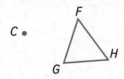

**4.** scale factor: $\frac{1}{5}$

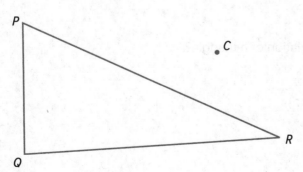

**MP** **Use Math Tools** For each figure from Exercise 3, measure the given side lengths in millimeters. Complete the table.

**5.**

| Figure | Side Lengths (mm) | | |
|--------|-----|-----|-----|
| | FG | GH | HF |
| △FGH | | | |

**6.**

| Figure | Side Lengths (mm) | | |
|--------|-----|-----|-----|
| | F'G' | G'H' | H'F' |
| △F'G'H' | | | |

**7.** What is the ratio of side *FG* to side *F'G'*? _____

**8.** What is the ratio of side *GH* to side *G'H'*? _____

**9.** What is the ratio of side *HF* to side *H'F'*? _____

**10.** Measure the angles of △FGH and △F'G'H' in Exercise 3 using a protractor.

Describe the relationship between the corresponding angles.

_____

_____

| Angle Measure (°) | | |
|-----|-----|-----|
| ∠F | ∠G | ∠H |
| | | |
| ∠F' | ∠G' | ∠H' |
| | | |

On Your Own

## Create

**11.** **MP** **Reason Inductively** Based on the Activities and Exercises, write a conjecture about the effects of a dilation on the sides and angles of a triangle.

_____

_____

_____

**12.** **inquiry** WHAT are the results of a dilation of a triangle?

_____

_____

# Lesson 4
# Dilations

## Vocabulary Start-Up

A dilation uses a scale factor to enlarge or reduce a figure.
**Scan the lesson and complete the graphic organizer.**

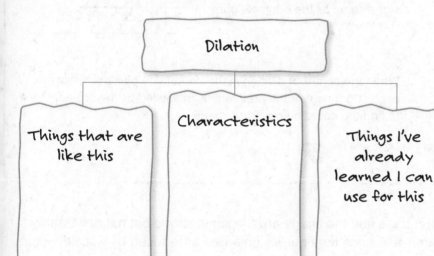

Dilation

Things that are like this

Characteristics

Things I've already learned I can use for this

 **Essential Question**

HOW can we best show or describe the change in position of a figure?

**Math Symbols**

$(x, y) \longrightarrow (kx, ky)$

 **Common Core State Standards**

**Content Standards**
8.G.3

**MP Mathematical Practices**
1, 3, 4

## Real-World Link

**Photography** Necie wants to insert a photo of her dog on her blog. The current size of the photo is 480 pixels by 640 pixels.

1. Suppose she wants to reduce the photo to 120 pixels by 160 pixels. Compare and contrast the original photo and the reduction. _____

2. What is the scale factor from the original to the reduction? _____

---

**Which MP Mathematical Practices did you use?**
**Shade the circle(s) that applies.**

① Persevere with Problems
② Reason Abstractly
③ Construct an Argument
④ Model with Mathematics

⑤ Use Math Tools
⑥ Attend to Precision
⑦ Make Use of Structure
⑧ Use Repeated Reasoning

# Dilations in the Coordinate Plane

**Words**   A dilation with a scale factor of $k$ will be:

- an enlargement, or an image larger than the original, if $k > 1$,

- a reduction, or an image smaller than the original, if $0 < k < 1$,

- the same as the original figure if $k = 1$.

**Model**

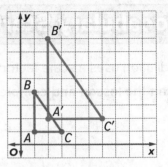

When the center of dilation in the coordinate plane is the origin, each coordinate of the preimage is multiplied by the scale factor $k$ to find the coordinates of the image.

**Symbols**   $(x, y) \longrightarrow (kx, ky)$

The preimage and the image are the same shape but not necessarily the same size since the figure is enlarged or reduced by a scale factor.

## Example

Watch   Tutor

**1.** **A triangle has vertices $A(0, 0)$, $B(8, 0)$, and $C(3, -2)$. Find the coordinates of the triangle after a dilation with a scale factor of 4.**

The dilation is $(x, y) \longrightarrow (4x, 4y)$. Multiply the coordinates of each vertex by 4.

$A(0, 0) \longrightarrow (4 \cdot 0, \ 4 \cdot 0) \longrightarrow (0, 0)$
$B(8, 0) \longrightarrow (4 \cdot 8, \ 4 \cdot 0) \longrightarrow (32, 0)$
$C(3, -2) \longrightarrow [4 \cdot 3, \ 4 \cdot (-2)] \longrightarrow (12, -8)$

So, the coordinates after the dilation are $A'(0, 0)$, $B'(32, 0)$, and $C'(12, -8)$.

**Got it?** Do this problem to find out.

Show your work.

a. _____

a. A figure has vertices $W(-2, 4)$, $X(1, 4)$, $Y(3, -1)$, and $Z(-3, -1)$. Find the coordinates of the figure after a dilation with a scale factor of 2.

Work Zone

# Example

Tutor

**2.** A figure has vertices $J(3, 8)$, $K(10, 6)$, and $L(8, 2)$. Graph the figure and the image of the figure after a dilation with a scale factor of $\frac{1}{2}$.

The dilation is $(x, y) \rightarrow \left(\frac{1}{2}x, \frac{1}{2}y\right)$.
Multiply the coordinates of each vertex by $\frac{1}{2}$. Then graph both figures on the coordinate plane.

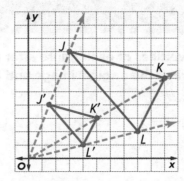

$J(3, 8) \rightarrow \left(\frac{1}{2} \cdot 3, \frac{1}{2} \cdot 8\right) \rightarrow J'\left(\frac{3}{2}, 4\right)$

$K(10, 6) \rightarrow \left(\frac{1}{2} \cdot 10, \frac{1}{2} \cdot 6\right) \rightarrow K'(5, 3)$

$L(8, 2) \rightarrow \left(\frac{1}{2} \cdot 8, \frac{1}{2} \cdot 2\right) \rightarrow L'(4, 1)$

**Check** Draw lines throught the origin and each of the vertices of the original figure. The vertices of the dilation should lie on those same lines. ✓

**STOP and Reflect**

Explain below how you can determine if a dilation is a reduction or an enlargement based on the scale factor.

## Got it? Do these problems to find out.

**b.** A figure has vertices $F(-1, 1)$, $G(1, 1)$, $H(2, -1)$, and $I(-1, -1)$. Graph the figure and the image of the figure after a dilation with a scale factor of 3.

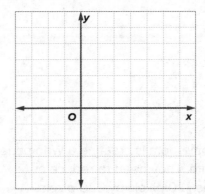

# Example

Tutor

**3.** Through a microscope, the image of a grain of sand with a 0.25-millimeter diameter appears to have a diameter of 11.25 millimeters. What is the scale factor of the dilation?

Write a ratio comparing the diameters of the two images.

$$\frac{\text{diameter in dilation}}{\text{diameter in original}} = \frac{11.25}{0.25}$$

$$= 45$$

So, the scale factor of the dilation is 45.

**Got it?** Do this problem to find out.

c. _____

c. Lucas wants to enlarge a 3- by 5-inch photo to a $7\frac{1}{2}$- by $12\frac{1}{2}$-inch photo. What is the scale factor of the dilation?

# Guided Practice

Check

**Find the coordinates of the vertices of each figure after a dilation with the given scale factor _k_. Then graph the original image and the dilation.**
(Examples 1 and 2)

**1.** $A(3, 5)$, $B(0, 4)$, $C(-2, -2)$; $k = 2$

_____

Show your work.

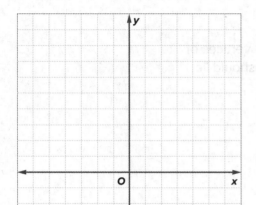

**2.** $J(0, -4)$, $K(0, 6)$, $L(4, 4)$, $M(4, 2)$; $k = \frac{1}{4}$

_____

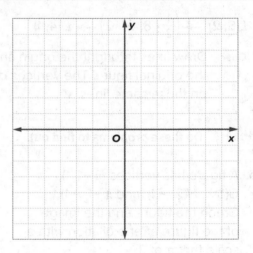

**3.** **STEM** Mrs. Bowen's homeroom is creating a Web page for their school's Intranet site. They need to reduce a scanned photograph so it is 720 pixels by 320 pixels. If the scanned photograph is 1,080 pixels by 480 pixels, what is the scale factor of the dilation? (Example 3) _____

**4.** **Building on the Essential Question** How are dilations similar to scale drawings?

_____

_____

_____

**Rate Yourself!**

☐ I understand how to dilate a figure.

▶▶ Great! You're ready to move on!

☐ I still have some questions about how to dilate a figure.

📖 No Problem! Go online to access a Personal Tutor. | Tutor

**FOLDABLES** Time to update your Foldable!

Name _____ My Homework _____

**Find the coordinates of the vertices of each figure after a dilation with the given scale factor k. Then graph the original image and the dilation.** (Examples 1 and 2)

**1** $C(1, 4)$, $A(2, 2)$, $T(5, 5)$; $k = 2$

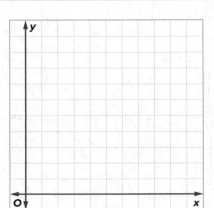

2. $R(1, 1)$, $S(1, 7)$, $T(5, 7)$, $U(5, 1)$; $k = \frac{3}{4}$

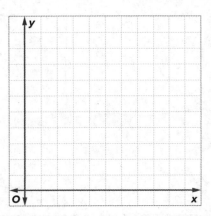

3. A graphic designer created a logo on $8\frac{1}{2}$- by 11-inch paper. In order to be placed on a business card, the logo needs to be $1\frac{7}{10}$ inches by $2\frac{1}{5}$ inches. What is the scale factor of the dilation? (Example 3)

_____

_____

4. Darian wants to build a regulation-size pool table that is 9 feet in length. The plans he ordered are 18 by 36 inches. What is the scale factor of the dilation he must use to build the regulation pool table? (Example 3)

_____

_____

**5** A triangle has vertices $A(-2, 3)$, $B(0, 0)$, and $C(1, 1)$.

    **a.** Find the coordinates of the triangle if it is reflected over the x-axis, then dilated by a scale factor of 3.

_____

    **b.** Find the coordinates if the original triangle is dilated by a scale factor of 3, then reflected over the x-axis.

_____

    **c.** Are the two transformations commutative? Explain.

_____

_____

6. **MP Model with Mathematics** In each part of the graphic organizer, sketch an image of pentagon *MNOPQ* after a dilation within the given parameters.

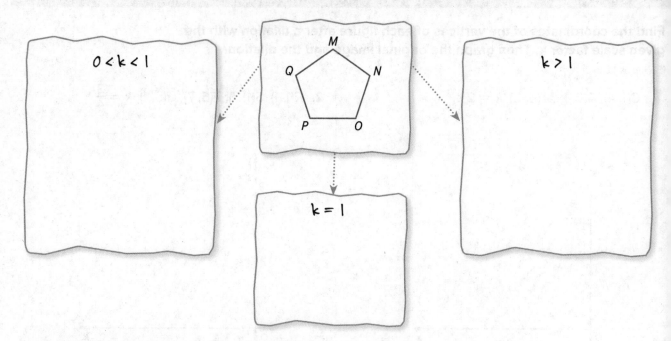

## H.O.T. Problems Higher Order Thinking

7. **MP Make a Conjecture** A figure has a vertex at the point $(-4, -6)$. The figure is dilated with the center at the origin with a scale factor of 5. The resulting image is then dilated with a scale factor of $\frac{3}{5}$.

   **a.** What are the coordinates of the vertex in the final image? _____

   **b.** How do they compare with those of the original image?

   _____

   **c.** Can you predict the scale factor of a compound dilation? Explain.

   _____

   _____

8. **MP Persevere with Problems** The coordinates of two triangles are shown in the table. Is $\triangle WXY$ a dilation of $\triangle ABC$? Explain.

| WXY | | ABC | |
|---|---|---|---|
| W | $(a, b)$ | A | $(4a, 2b)$ |
| X | $(a, c)$ | B | $(4a, 2c)$ |
| Y | $(d, b)$ | C | $(4d, 2b)$ |

_____

_____

_____

9. **MP Persevere with Problems** The algebraic representation of a dilation is $(x, y) \rightarrow \left(\frac{1}{a}x, \frac{1}{a}y\right)$. If the dilation is an enlargement, give three possible

values of $a$. _____

# Extra Practice

**Find the coordinates of the vertices of each figure after a dilation with the given scale factor k. Then graph the original image and the dilation.**

**10.** $R(5, 5)$, $S(5, 10)$, $T(10, 10)$, $U(10, 5)$; $k = \frac{2}{5}$

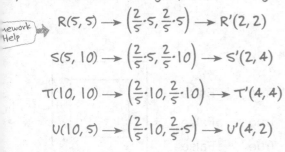

Multiply each coordinate in each ordered pair by scale factor. Then graph the two figures.

*(Homework Help)*

$$R(5, 5) \longrightarrow \left(\frac{2}{5} \cdot 5, \frac{2}{5} \cdot 5\right) \longrightarrow R'(2, 2)$$

$$S(5, 10) \longrightarrow \left(\frac{2}{5} \cdot 5, \frac{2}{5} \cdot 10\right) \longrightarrow S'(2, 4)$$

$$T(10, 10) \longrightarrow \left(\frac{2}{5} \cdot 10, \frac{2}{5} \cdot 10\right) \longrightarrow T'(4, 4)$$

$$U(10, 5) \longrightarrow \left(\frac{2}{5} \cdot 10, \frac{2}{5} \cdot 5\right) \longrightarrow U'(4, 2)$$

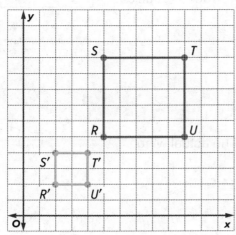

**11.** $V(-3, 4)$, $X(-2, 0)$, $W(1, 2)$; $k = 3$

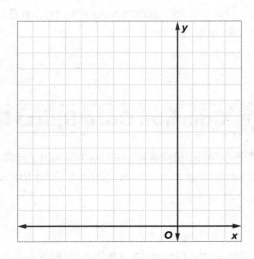

**12.** To place a picture in his class newsletter, Joaquin must reduce the picture by a scale factor of 0.3. Find the dimensions of the reduced picture if the original is 15 centimeters wide and 10 centimeters high.

**13.** **(MP) Multiple Representations** Triangle *XYZ* has vertices $X(0, 0)$, $Y(3, 1)$ and $Z(2, 3)$.
   **a. Numbers** Find the coordinates of the image of $\triangle XYZ$ after a dilation with a scale factor of $-2$.

   **b. Algebra** Graph $\triangle XYZ$ and the image on the coordinate plane.

   **c. Words** Describe the locations of $\triangle XYZ$ and $\triangle X'Y'Z'$ using transformations.

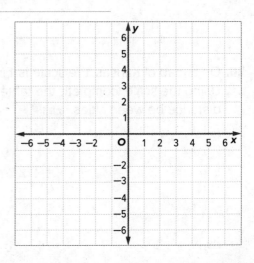

**14.** Triangle *RST* is dilated so that the image of point *T* is *T'*(6, 6). Draw triangle *R'S'T'*.

What is the scale factor of the dilation? Does the dilation represent an enlargement or a reduction? Explain how you found your answer.

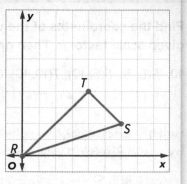

**15.** Squares *A* and *B* are related by a dilation. Determine if each statement is true or false.

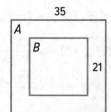

a. The scale factor from figure *A* to figure *B* is $\frac{3}{5}$.　☐ True ☐ False

b. The scale factor from figure *B* to figure *A* is $\frac{5}{3}$.　☐ True ☐ False

c. The dilation from figure *A* to figure *B* is an enlargement.　☐ True ☐ False

## (CCSS) Common Core Spiral Review

**16.** A model airplane is built with a wing span of 18 inches. The actual wing span of the airplane is 90 feet. Find the scale. **7.G.1**

**Find the scale factor for each scale. 7.G.1**

**17.** 6 in. = 10 ft _____

**18.** 4 cm = 2.5 mm _____

**19.** 5 ft = 15 yd _____

**20.** On a map of Kansas, the scale is 1 in. = 50 miles. Using the scale, complete the table showing the distance between cities. **7.G.1**

| Cities | Map | Actual |
|---|---|---|
| Wichita to Topeka | $2\frac{3}{4}$ in. | |
| Salina to Kansas City | | 150 mi |

# 21ST CENTURY CAREER
## in Computer Animation

## Computer Animator

Have you ever wondered how they make animated movies look so realistic? Computer animators use computer technology and apply their artistic skills to make inanimate objects come alive. If you are interested in computer animation, you should practice drawing, study human and animal movement, and take math classes every year in high school. Tony DeRose, a computer scientist at an animation studio said, "Trigonometry helps rotate and move characters, algebra creates the special effects that make images shine and sparkle, and calculus helps light up a scene."

**College & Career**
R E A D I N E S S

Explore college and careers at ccr.mcgraw-hill.com

## Is This the Career for You?

Are you interested in a career as a computer animator? Take some of the following courses in high school.

◆ 2-D Animation
◆ Algebra
◆ Calculus
◆ Trigonometry

Turn the page to find out how math relates to a career in Computer Animation.

## MP An Animation Sensation

**Use Figures 1–3 to solve each problem.**

1. In Figure 1, the car is translated 8 units left and 5 units down so that it appears to be moving. What are the coordinates of A' and B' after the translation? _____

2. In Figure 1, the car is translated so that A' has coordinates (−7, 2). Describe the translation as an ordered pair. Then find the coordinates of point B'. _____

3. In Figure 1, the car is reflected over the x-axis in order to make its reflection appear in a pond. What are the coordinates of A' and B' after the reflection? _____

4. In Figure 2, the artist uses rotation to show the girl's golf swing. Describe the coordinates of G' if the golf club is rotated 90° clockwise about point H. _____

5. The character in Figure 3 is enlarged by a scale factor of $\frac{5}{2}$. What are the coordinates of Q' and R' after the dilation? _____

6. The character in Figure 3 is reduced in size by a scale factor of $\frac{2}{3}$. What is the number of units between S' and T', the width of the character's face, after the dilation? _____

Figure 1

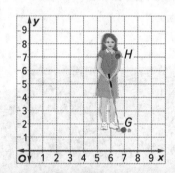

Figure 2

Figure 3

## MP Career Project

It's time to update your career portfolio! Choose a movie that was completely or partially computer animated. Use the Internet to research how technology was used to create the scenes in the movie. Describe any challenges that the computer animators faced.

_____

_____

_____

_____

_____

What are some short term goals you need to achieve to become a computer animator?

- _____
- _____
- _____
- _____
- _____

# Power Up! Performance Task

## Yearbook Layout

Students in Mrs. Johnson's fifth period class are experimenting with different page layouts on a computer screen. The coordinate grid at the right represents one page of a two-page spread. One photo is already placed on the page.

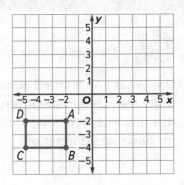

**Write your answers on another piece of paper. Show all of your work to receive full credit.**

### Part A

A second photo will be added by reflecting the original photo across the x-axis. Use a separate coordinate plane to draw and label the second photo. List the coordinates of the second photo.

### Part B

Mrs. Johnson wants the students to rotate the second photo 90° clockwise about the origin, then translate it two units down and one unit to the right to place a third photo. Draw and label the third photo on your coordinate plane. List the coordinates of the third photo.

### Part C

DeShawn placed a square photo in the center of the screen as shown. Dilate the photo so that it results in the picture taking up the whole screen. What is the scale factor? Label each point with the new coordinates.

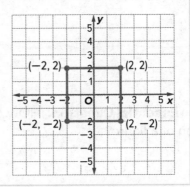

# Reflect

 **Answering the Essential Question**

Use what you learned about transformations to complete the graphic organizer. Determine if you would demonstrate each transformation by using words, symbols, or models. Then give an example using your method for each transformation.

translation

reflection

**Essential Question**

HOW can we best show or describe the change in position of a figure?

rotation

dilation

**Answer the Essential Question.** HOW can we best show or describe the change in position of a figure?

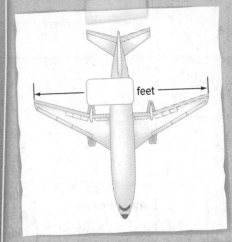

# Chapter 7
# Congruence and Similarity

 **Essential Question**

HOW can you determine congruence and similarity?

 **Common Core State Standards**

**Content Standards**
8.G.1, 8.G.1a, 8.G.1b, 8.G.2, 8.G.4, 8.G.5, 8.EE.6

 **Mathematical Practices**
1, 2, 3, 4, 5, 7

 **Math in the Real World**

**Models** The wingspan of a model of a 737 commercial aircraft is 6.75 inches. The scale for the model is
1 inch = 200 inches.

Use the scale to find the wingspan in inches of the actual 737 aircraft. Then convert the inches to feet.

feet

 **FOLDABLES** **Study Organizer**

 **1** Cut out the Foldable on page FL7 of this book.

**2** Place your Foldable on page 580.

**3** Use the Foldable throughout this chapter to help you learn about congruence and similarity.

## Vocabulary

composition of transformations

corresponding parts

indirect measurement

scale factor

similar

similar polygons

## Study Skill: Use a Web

**Use a Web**  A *web* can help you understand how math concepts are related to each other. To make a web, write the major topic in the center of a piece of paper. Then, draw "arms" from the center for as many categories as you need.

Here is a partial web for the major topic of triangles. Complete the web by adding descriptions for the classifications by sides. Then add the classifications by angles.

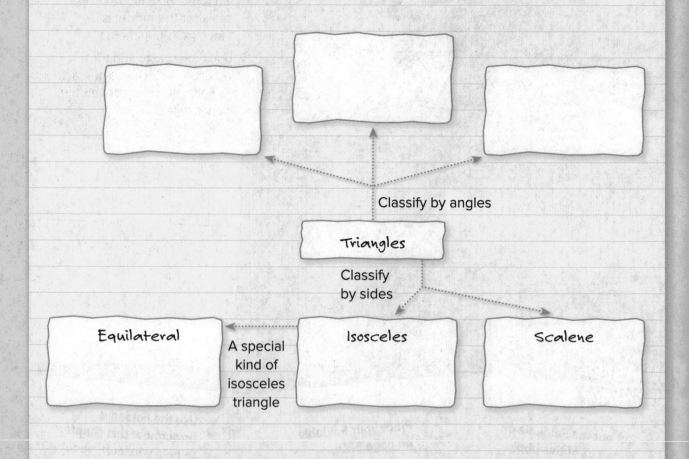

# What Do You Already Know?

List three things you already know about congruence and similarity in the first section. Then list three things you would like to learn about congruence and similarity in the second section.

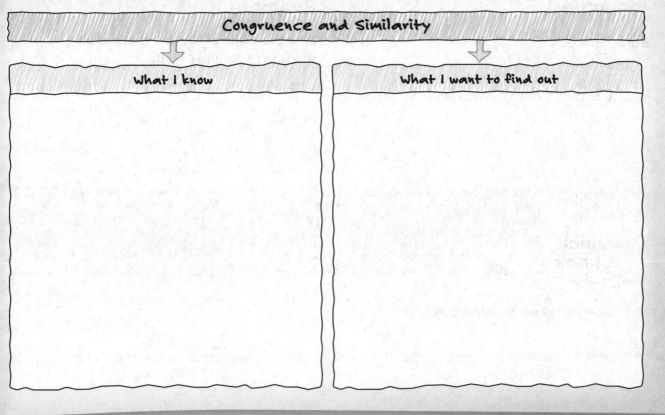

Congruence and Similarity

What I know

What I want to find out

# When Will You Use This?

Here are a few examples of how triangles are used in the real world.

**Activity 1** Have you ever flown a kite? How much string did you have on the spool?

_____

_____

**Activity 2** Go online at **connectED.mcgraw-hill.com** to read the graphic novel *Up, Up, and Away*. What is the length of the rope

that is attached to the parasailer? _____

Jasmine and Jacob in
Up, Up, and Away

Try the Quick Check below.
Or, take the Online Readiness Quiz.

 Check ✓

**CCSS** **Quick Review**

**Common Core Review** 7.RP.2, 8.EE.5

## Example 1

Solve $\frac{w}{12} = \frac{5}{6}$.

$\frac{w}{12} = \frac{5}{6}$     Write the proportion.

$6 \times w = 12 \times 5$     Find cross products.

$6w = 60$     Simplify.

$w = 10$     Division Property of Equality

## Example 2

**Find the slope of the line that passes through (3, 8) and (−1, 0).**

$m = \frac{y_2 - y_1}{x_2 - x_1}$     Slope formula

$m = \frac{0 - 8}{-1 - 3}$     $(x_1, y_1) = (3, 8); (x_2, y_2) = (-1, 0)$

$m = \frac{-8}{-4}$ or 2     Simplify.

## Quick Check

**Proportions** Solve each proportion.

**1.** $\frac{x}{15} = \frac{7}{30}$ _____

**2.** $\frac{4}{9} = \frac{14}{y}$ _____

**3.** $\frac{12}{z} = \frac{30}{37}$ _____

 Show your work.

**4.** $\frac{8}{15} = \frac{m}{21}$ _____

**5.** $\frac{n}{5} = \frac{18}{45}$ _____

**6.** $\frac{3}{7} = \frac{21}{p}$ _____

**Find Slope** Find the slope of the line that passes through each pair of points.

**7.** (−1, 1), (−3, 7) _____

**8.** (2, 0), (0, 2) _____

**9.** (−6, −1), (−3, 4) _____

**How Did You Do?**

**Which problems did you answer correctly in the Quick Check? Shade those exercise numbers below.**

① ② ③ ④ ⑤ ⑥ ⑦ ⑧ ⑨

# Inquiry Lab
## Composition of Transformations

 **Inquiry** HOW does a combination of transformations differ from a single transformation? How are they the same?

 **CCSS** **Content Standards** 8.G.2

 **MP** **Mathematical Practices** 1, 3

Graphic artists often use several transformations to create designs. When a transformation is applied to a figure and then another transformation is applied to the image, the result is called a **composition of transformations.**

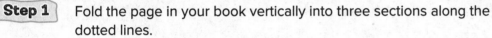

## Hands-On Activity 1

**Step 1** Fold the page in your book vertically into three sections along the dotted lines.

**Step 2** Draw the reflection of the arrow over the fold in the middle section.

**Step 3** Draw a reflection of the 2ⁿᵈ arrow over the fold in the right-hand section.

**Step 4** Repeat Steps 2 and 3 with the pentagon.

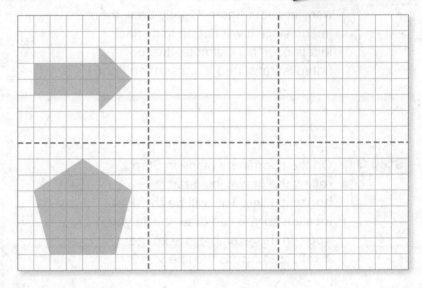

How are the original figures and the final figures related?

_____

_____

Would the final images be the same as the original figure if the second reflection was reflected over the horizontal line? Explain.

_____

_____

# Hands-On Activity 2

In this Activity, you will use a translation and a reflection to create a decorative border.

**Step 1**   Draw a figure on the coordinate plane shown, close to the origin.

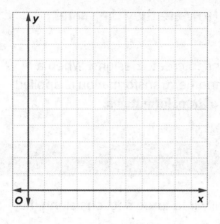

**Step 2**   On the coordinate plane in Step 1, translate your figure. Lightly draw the image since it will not be in its final location. In this example, the red figure is translated 2 units to the right.

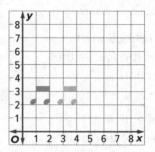

**Step 3**   On the coordinate plane in Step 1, reflect the drawn image across a horizontal line. This will be the final location so you can draw this in your book. In this example, the image is reflected across the line $y = 2$.

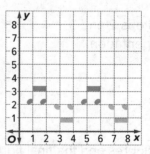

**Step 4**   Repeat the process to create your border.

How are the size and shape of the original figure related to the size and shape of the images?

_____

Suppose you wanted your border to run up the side of the page instead of across the bottom of the page. Describe what transformations you might use

to do this. _____

_____

# Investigate

Collaborate

**Work with a partner. Describe the transformations combined to create the outlined patterns shown in Exercises 1–4.**

**1.**

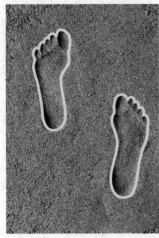

**2.**

**3.**

**4.**

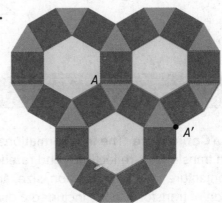

**5.** Draw a figure on the coordinate plane shown. Use a reflection and a rotation to create a logo for a company.

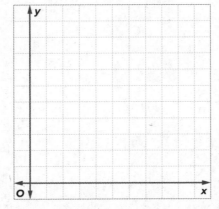

## Analyze and Reflect

Collaborate

In some cases, a composition of transformations is the same as a single transformation. Draw the composition of transformations described. Then identify the single transformation that would produce the same image as each composition.

6. $\overline{AB}$ is reflected across the $y$-axis, then reflected across the $x$-axis.

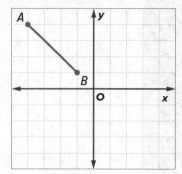

_____

_____

7. $\triangle XYZ$ is reflected across line $m$ and then reflected across line $n$.

_____

## Create

On Your Own

8. **MP** **Make a Conjecture** The transformations in the Activities and Exercises have been translations, reflections, and rotations which preserve distance. Make a conjecture about the position, size, and shape of a figure if a composition of transformations included a dilation.

_____

_____

9. **inquiry** HOW does a combination of transformations differ from a single transformation? How are they the same?

_____

_____

_____

_____

_____

# Congruence and Transformations

 ## Real-World Link

**Braille** The letter R in the Braille alphabet consists of four large dots and 2 smaller dots in the pattern shown. Circle the letter with the same shape as the letter R.

R:

 **Essential Question**

HOW can you determine congruence and similarity?

**Common Core State Standards**

Content Standards
8.G.1, 8.G.1a, 8.G.1b, 8.G.2

MP **Mathematical Practices**
1, 3, 4

 **How can you determine whether two figures are the same size and shape?**

**Step 1** Copy the figure shown on tracing paper two times. Cut out both figures. Label the figures A and B.

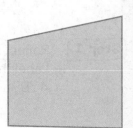

**Step 2** Place Figure B on top of Figure A. Are the side lengths the same? the angle measures?

Are the figures the same size and shape? _____

**Step 3** Translate Figure B up and over on your desk. How can you move Figure A on top of Figure B so all sides and angles match?

_____

_____

**Step 4** Flip Figure B over. How can you move Figure A on top of Figure B so all sides and angles match?

 **Which MP Mathematical Practices did you use?**
**Shade the circle(s) that applies.**

① Persevere with Problems
② Reason Abstractly
③ Construct an Argument
④ Model with Mathematics
⑤ Use Math Tools
⑥ Attend to Precision
⑦ Make Use of Structure
⑧ Use Repeated Reasoning

# Identify Congruence

On the previous page, you matched Figure *A* to Figure *B* by a translation and a reflection. Two figures are congruent if the second can be obtained from the first by a series of rotations, reflections, and/or translations.

## Examples

**Determine if the two figures are congruent by using transformations. Explain your reasoning.**

**1.**

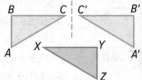

| **Step 1** | Reflect △*ABC* over a vertical line. Label the vertices of the image *A'*, *B'*, and *C'*. |

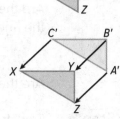

| **Step 2** | Translate △*A'B'C'* until all sides and angles match △*XYZ*. |

So, the two triangles are congruent because a reflection followed by a translation will map △*ABC* onto △*ZYX*.

**2.**

Reflect the red figure over a vertical line.

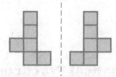

Even if the reflected figure is translated up and over, it will not match the green figure exactly. The two figures are not congruent.

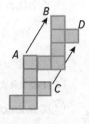

**Got it?** Do these problems to find out.

a.

A
B

b.

a. _____

b. _____

# Determine the Transformations

If you have two congruent figures, you can determine the transformation, or series of transformations, that maps one figure onto the other by analyzing the orientation or relative position of the figures.

Show your work.

| Translation | Reflection | Rotation |
|---|---|---|
| • length is the same<br>• orientation is the same | • length is the same<br>• orientation is reversed | • length is the same<br>• orientation is the same |
|  | | |

## Orientation

The order in which the vertices of a figure are named determines the figure's orientation. In the reflection shown, the vertices of the preimage are named in a clockwise direction, but the vertices of the image are named in a counterclockwise direction. The orientation has been reversed.

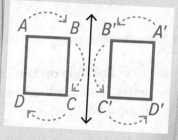

## Example

Tutor

**3.** Ms. Martinez created the logo shown. What transformations did she use if the letter "d" is the preimage and the letter "p" is the image? Are the two figures congruent?

iamond
lumbing

**Step 1** Start with the preimage. Rotate the letter "d" 180° about point *A*.

**Step 2** Translate the new image up.

Ms. Martinez used a rotation and translation to create the logo. The letters are congruent because images produced by a rotation and translation have the same shape and size.

**Got it?** Do this problem to find out.

c. _____

c. What transformations could be used if the letter "W" is the preimage and the letter "M" is the image in the logo shown? Are the two figures congruent? Explain.

Check ✓

**Determine if the two figures are congruent by using transformations. Explain your reasoning.** (Examples 1 and 2)

1.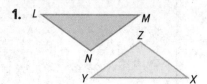

2.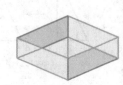

_____

_____

_____

_____

_____

_____

3. The Boyd Box Company uses the logo shown. What transformations could be used if the red trapezoid is the preimage and the blue trapezoid is the image? Are the two figures congruent? Explain. (Example 3)

_____

_____

_____

4.  **Building on the Essential Question** Why do translations, reflections, and rotations create congruent images?

_____

_____

_____

_____

**Rate Yourself!**

How confident are you about the relationship between congruence and transformations? Check the box that applies.

For more help, go online to access a Personal Tutor.

Tutor

**FOLDABLES** Time to update your Foldable!

# Independent Practice

Go online for Step-by-Step Solutions
eHelp

**Determine if the two figures are congruent by using transformations. Explain your reasoning.** (Examples 1 and 2)

**1**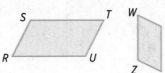

_____

_____

_____

**2.**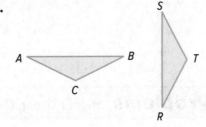

_____

_____

_____

**3** Nilda purchased some custom printed stationery with her initials. What transformations could be used if the letter "Z" is the preimage and the letter "N" is the image in the design shown? Are the two figures

congruent? Explain. (Example 3) _____

_____

_____

**4.** **MP** **Multiple Representations** One way to identify congruent triangles is to prove their matching sides have the same measure. Triangle *CDE* has vertices at (1, 4), (1, 1), and (5, 1).

a. **Graphs** Graph △*CDE*.

b. **Numbers** Find the lengths of the sides of △*CDE*.

_____

c. **Geometry** Reflect △*CDE* over the *y*-axis, then translate it 2 units left. Label the vertices of the image *C'D'E'*. Write the coordinates of △*C'D'E'* below.

_____

d. **Numbers** Find the lengths of the sides of △*C'D'E'*.

_____

e. **Words** Are the two triangles congruent? Justify your response.

_____

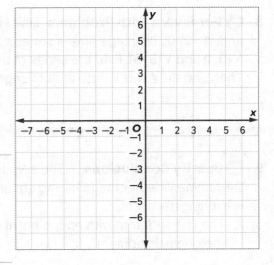

**5.** Graph △*GHJ* with vertices at *G*(0, 1), *H*(4, 0), and *J*(4, 1). Then graph the image of the triangle after a translation of 3 units up followed by a reflection over the *y*-axis. Find the lengths of each side of the preimage and the image. Then determine if the two figures are congruent.

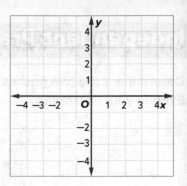

---

## 🔥 H.O.T. Problems  Higher Order Thinking

**6.** **MP Model with Mathematics** Create a design in the space at the right, using a series of transformations that produce congruent figures. Exchange designs with a classmate and determine what transformations were used to create their design.

Show your work.

**7.** **MP Persevere with Problems** Triangle *A'B'C'* has vertices *A'*(−4, 5), *B'*(−1, 4), and *C'*(−2, 0). Triangle *ABC* was rotated 90° in a clockwise direction about the origin, translated 2 units up, and reflected over the *y*-axis. What were the coordinates of the vertices of triangle *ABC*?

**8.** **MP Persevere with Problems** Line segment *XY* has endpoints at *X*(3, 1) and *Y*(−2, 0). Its image after a series of transformations has endpoints at *X'*(0, 1) and *Y'*(5, 0). Find the series of transformations that maps $\overline{XY}$ onto $\overline{X'Y'}$. Then find the exact length of both segments.

**9.** **MP Justify Conclusions** A line segment has endpoints at (*a*, *b*) and (*c*, *d*). Determine whether the following statements are *true* or *false*. Justify your reasoning.

**a.** The line segment with endpoints at (*a* + *x*, *b*) and (*c* + *x*, *d*) is congruent to the original segment.

**b.** The line segment with endpoints at $\left(\frac{2}{3}a, \frac{2}{3}b\right)$ and $\left(\frac{2}{3}c, \frac{2}{3}d\right)$ is congruent to the original segment.

# Extra Practice

**Determine if the two figures are congruent by using transformations. Explain your reasoning.**

**10.**

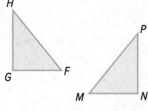

➡ The two figures are not congruent because no sequence of

transformations will map the green figure onto the red figure exactly.

**11.**

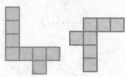

_____

_____

_____

**12.**

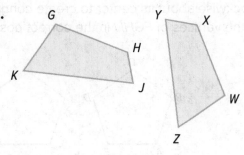

_____

_____

_____

_____

**13.** Simon is illustrating a graphic novel for a friend. He is using the two thought bubbles shown. What transformations did he use if Figure A is the preimage and Figure B is the image?

**14.** **MP** **Model with Mathematics** Graph $\triangle PQR$ with vertices at $P(0, 0)$, $Q(2, 0)$, and $R(0, 2)$. Then graph the image of the triangle after a reflection over the x-axis followed by a dilation with a scale factor of 2. Find the lengths of each side of the preimage and the image. Then determine if the two figures are congruent.

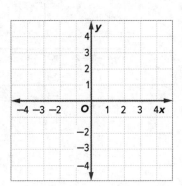

_____

_____

**15.** Gregory is creating a mosaic for art class. He started by using triangular tiles as shown.

Triangles *A* and *B* are congruent. Describe possible transformations he could have used if triangle *A* is the preimage and triangle *B* is the image?

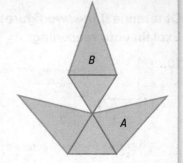

**16.** Pentagon *ABCDE* is reflected across the line shown and then rotated 72° clockwise about its center to create congruent pentagon *FGHIJ*. Label the vertices of *FGHIJ* in the correct positions on the image.

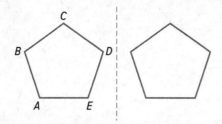

**Graph each figure with the given vertices and its image after the indicated transformation. Then give the coordinates of the final image.** 8.G.1, 8.G.3

**17.** $\overline{CD}$: C(−2, 4), D(0, 0); translation of 3 units right and 2 units down

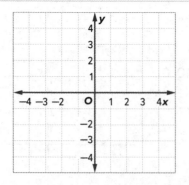

**18.** △*XYZ*: X(−1, 1), Y(3, 1), Z(1, 3); reflection over the y-axis

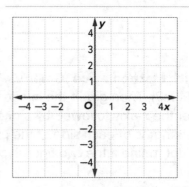

# Inquiry Lab
## Investigate Congruent Triangles

**Inquiry** WHICH three pairs of corresponding parts can be used to show that two triangles are congruent?

 **Content Standards** 8.G.2

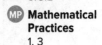 **Mathematical Practices** 1, 3

While driving past a bridge with his family, Henry noticed that the bridge truss was made up of congruent triangles.

## Hands-On Activity 1

In this Activity you will investigate whether it is possible to show that two triangles are congruent without showing that all six pairs of corresponding parts are congruent.

**Step 1** Copy the sides of the triangle shown onto a piece of patty paper and cut them out.

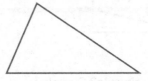

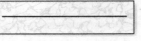

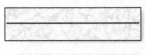

**Step 2** Arrange and tape the pieces together so that they form a triangle.

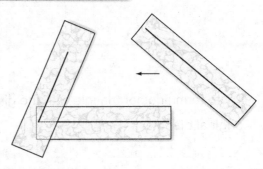

Is the triangle you formed congruent to the original triangle?

Explain. _____

_____

Rotate the triangle you formed 180°. Is the triangle congruent to the original

triangle? Explain. _____

_____

Try to form another triangle with the given sides. Is it congruent to the original

triangle? _____

# Hands-On Activity 2

**Step 1**  Draw a triangle on a piece of patty paper. Copy each angle of the triangle onto a separate piece of patty paper. Extend each side of each angle to the edge of the patty paper.

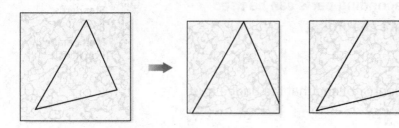 

**Step 2**  Arrange and tape the pieces together so that they form a triangle.

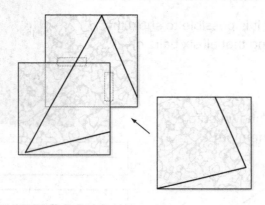

Is the triangle you formed congruent to the original triangle? Explain.

_____

_____

_____

Try to form another triangle with the given angles. Is it congruent to the original triangle? _____

_____

A *counterexample* disproves a statement by showing an example of when the statement is not true. Based on this activity, is the following statement true? If not, provide a counterexample.

*If the angles of one triangle are congruent to the angles of another triangle, the two triangles are congruent.*

_____

## Investigate

Collaborate

1. Draw a triangle on a piece of patty paper. Copy two sides of the triangle and the angle between them onto separate pieces of patty paper and cut them out. Arrange and tape pieces together so that the two sides are joined to form the rays of the angle. Connect the two rays to form a triangle.

   **a.** Is the triangle you formed congruent to the original triangle?

   Explain. _____

   _____

   **b.** Try to form another triangle with the given sides and angle. Is it

   congruent to the original triangle? _____

2. Determine if two triangles with the following congruent parts are congruent. If not, draw a counterexample.

| Various Parts | Congruent? | Counterexample |
|---|---|---|
| 3 angles | No | |
| 2 sides | | |
| 2 angles and 1 side | | |
| 2 angles and the side between the 2 angles | | |
| 2 angles | | |
| 3 sides | | |

## Analyze and Reflect

**3.** Based on Activity 1, can three pairs of congruent sides be used to show that

two triangles are congruent? _____

**4.** Based on Activity 2, can three pairs of congruent angles be used to show that

two triangles are congruent? _____

**5.** Based on Exercise 1, can two pairs of congruent sides and the pair of
congruent angles between them be used to show that two triangles are

congruent? _____

## Create

On Your Own

**6.** **MP** **Make a Conjecture** Use patty paper to investigate the relationship
between two triangles with the given information. Make a conjecture about
whether each of these cases can be used to show that two triangles are
congruent.

**Case 1** two pairs of congruent sides and a pair of congruent angles

not between them _____

_____

_____

**Case 2** two pairs of congruent angles and the pair of congruent sides

between them _____

_____

_____

**Case 3** two pairs of congruent angles and a pair of congruent sides

not between them _____

_____

_____

**7.** **inquiry** WHICH three pairs of corresponding parts can be used to show that
two triangles are congruent?

_____

_____

_____

# Congruence

 **Real-World Link**

**Crafts** Lauren is creating a quilt using the geometric pattern shown. She wants to make sure that all of the triangles in the pattern are the same shape and size.

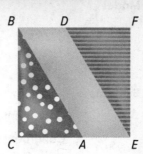

1. What would Lauren need to do to show the two triangles are congruent?

   _____

   _____

2. Complete the lists of the parts of △ABC and △DEF. Then draw lines between the corresponding parts of each triangle.

   $\overline{CB}$ _____ $\overline{BA}$ ∠BAC ∠ABC ∠ _____

   _____ $\overline{ED}$ _____ ∠ _____ ∠ _____ ∠EDF

3. Suppose you cut out the two triangles and laid one on top of the other so the parts of the same measures were matched up. What is true about the triangles?

   _____

 **Essential Question**

HOW can you determine congruence and similarity?

**Vocab**  **Vocabulary**

corresponding parts

**Math Symbols**
≅ is congruent to

**CCSS** **Common Core State Standards**

**Content Standards**
8.G.2

**MP** **Mathematical Practices**
1, 2, 3, 4

Which **MP** **Mathematical Practices** did you use?
Shade the circle(s) that applies.

① Persevere with Problems     ⑤ Use Math Tools

② Reason Abstractly     ⑥ Attend to Precision

③ Construct an Argument     ⑦ Make Use of Structure

④ Model with Mathematics     ⑧ Use Repeated Reasoning

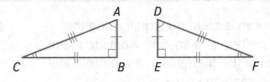

## Key Concept

## Corresponding Parts of Congruent Figures

*Work Zone*

**Congruence**

To indicate that sides are congruent, an equal number of tick marks is drawn on the corresponding sides. To show that angles are congruent, an equal number of arcs is drawn on the corresponding angles.

**Words**   If two figures are congruent, their corresponding sides are congruent and their corresponding angles are congruent.

**Model**

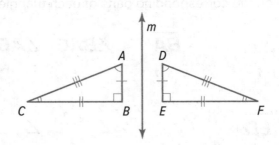

**Symbols**        $\triangle ABC \cong \triangle DEF$

Congruent Angles:  $\angle A \cong \angle D$; $\angle B \cong \angle E$; $\angle C \cong \angle F$

Congruent Sides:   $\overline{AB} \cong \overline{DE}$; $\overline{BC} \cong \overline{EF}$; $\overline{CA} \cong \overline{FD}$

In the figure below, the two triangles are congruent because $\triangle DEF$ is the image of $\triangle ABC$ reflected over line $m$. The notation $\triangle ABC \cong \triangle DEF$ is read *triangle ABC is congruent to triangle DEF*.

The parts of congruent figures that *match* or correspond, are called **corresponding parts**.

## Example

Tutor

1.  **Write congruence statements comparing the corresponding parts in the congruent triangles shown.**

    Use the matching arcs and tick marks to identify the corresponding parts.

    Corresponding angles:
    $\angle J \cong \angle G$, $\angle L \cong \angle I$, $\angle K \cong \angle H$

    Corresponding sides:
    $\overline{JK} \cong \overline{GH}$, $\overline{KL} \cong \overline{HI}$, $\overline{LJ} \cong \overline{IG}$

    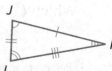

**Got it?** Do this problem to find out.

a.

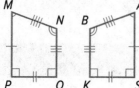

# Example

Tutor

**2.** Triangle *ABC* is congruent to △*XYZ*. Write congruence statements comparing the corresponding parts. Then determine which transformations map △*ABC* onto △*XYZ*.

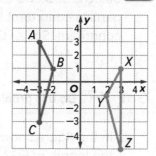

**STOP and Reflect**

When writing congruence statements, why is it important to match up corresponding points in the statement?

**Step 1** Analyze the figures to determine which angles and sides of the figures correspond.

Corresponding angles: ∠*A* ≅ ∠*X*, ∠*B* ≅ ∠*Y*, ∠*C* ≅ ∠*Z*
Corresponding sides: $\overline{AB}$ ≅ $\overline{XY}$, $\overline{BC}$ ≅ $\overline{YZ}$, $\overline{CA}$ ≅ $\overline{ZX}$

**Step 2** Determine any changes in the orientation of the triangles. The orientation is reversed so at least one of the transformations is a reflection. If you reflect △*ABC* over the *y*-axis and then translate it down 2 units, it coincides with △*XYZ*.

The transformations that map △*ABC* onto △*XYZ* consist of a reflection over the *y*-axis followed by a translation of 2 units down.

**Got it?** Do this problem to find out.

**b.** Parallelogram *WXYZ* is congruent to parallelogram *KLMN*. Write congruence statements comparing the corresponding parts. Then determine which transformation(s) map parallelogram *WXYZ* onto parallelogram *KLMN*.

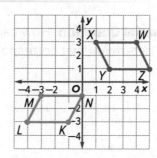

## Find Missing Measures

You can use properties of congruent figures to find the missing measures of angles and sides in a figure.

### Example

**3.** Miley is using a brace to support a tabletop. In the figure, △*BCE* ≅ *DFG*. If *m∠CEB* = 50°, what is the measure of ∠*FGD*?

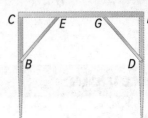

Since ∠*CEB* and ∠*FGD* are corresponding parts in congruent figures, they are congruent. So, ∠*FGD* measures 50°.

**Got it?** Do this problem to find out.

c. _____

**c.** In the figure shown above, the length of $\overline{CE}$ is 2 feet. What is the length of $\overline{FG}$?

## Guided Practice

**1.** Triangle *RST* is congruent to △*UVW*. Write congruence statements comparing the corresponding parts. Then determine which transformation(s) map △*RST* onto △*UVW*. (Examples 1 and 2)

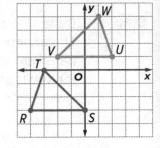

_____

_____

_____

**2.** In the table design shown in Example 3, suppose *BE* = 18 inches. What is *DG*? (Example 3)

_____

**3.**  **Building on the Essential Question** How can the coordinate plane help you determine that corresponding sides are congruent? _____

_____

_____

**Rate Yourself!**

How confident are you about congruence? Check the box that applies.

For more help, go online to access a Personal Tutor.

**FOLDABLES** Time to update your Foldable!

Name _____ My Homework _____

# Independent Practice

Go online for Step-by-Step Solutions

**Write congruence statements comparing the corresponding parts in each set of congruent figures.** (Example 1)

1.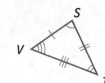

_____

_____

2.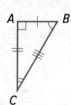

_____

_____

3. Parallelograms *UVWX* and *HJIK* are congruent. Write congruence statements comparing the corresponding parts. Then determine which transformation(s) map parallelogram *UVWX* onto parallelogram *HJIK*. (Example 2)

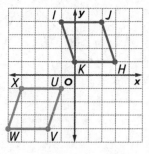

_____

_____

_____

4. In the umbrella shown at the right, △*JLK* ≅ △*NLM*. (Example 3)

   a. If *m∠JKL* = 66°, then *m∠NML* = _____.

   b. If *MN* = 15 inches, then *KJ* = _____ inches.

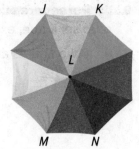

5. **Reason Abstractly** In the figure, △*ABC* ≅ △*EBD*.

   a. On the figure, draw arc and tic marks to identify the corresponding parts.

   b. Find the value of *x*.

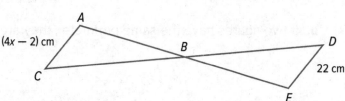

_____

6. In the figure at the right, △*EFG* ≅ △*LMN*. Find the value of *x*. Then describe the transformations that map △*EFG* onto △*LMN*.

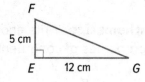 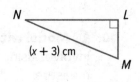

_____

_____

7. **MP** **Make a Conjecture** Hexagon *ABCDEF* has six congruent sides.

   a. Draw $\overline{CA}$, $\overline{CF}$, and $\overline{CE}$.

   b. How many triangles were formed? _____

   c. Make a conjecture about which triangles are congruent. Test your conjecture by measuring the sides and angles of the triangles.

   _____

   _____

## 🔥 H.O.T. Problems  Higher Order Thinking

8. **MP** **Find the Error** Mandar is making a congruence statement for the congruent triangles shown. Find his mistake and correct it.

   Triangle ABC is congruent to triangle DEF.

   _____

   _____

9. **MP** **Persevere with Problems** Determine whether each statement is *true* or *false*. If true, explain your reasoning. If false, give a counterexample.

   a. If two figures are congruent, their perimeters are equal.

   _____

   _____

   _____

   b. If two figures have the same perimeter, they are congruent.

   _____

   _____

   _____

10. **MP** **Model with Mathematics** Write and solve a real-world problem that involves using the properties of congruent figures to find a missing measure.

   _____

   _____

# Extra Practice

**Write congruence statements comparing the corresponding parts in each set of congruent figures.**

**11.**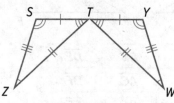

Use the matching arcs and tick marks to identify the corresponding parts.

Corresponding angles:

$\angle S \cong \angle Y$, $\angle STZ \cong \angle YTW$, $\angle Z \cong \angle W$

Corresponding sides:

$\overline{SZ} \cong \overline{YW}$, $\overline{ZT} \cong \overline{WT}$, $\overline{TS} \cong \overline{TY}$

**12.**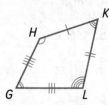

_____

_____

**13.** Quadrilaterals *KLMN* and *FGHJ* are congruent. Write congruence statements comparing the corresponding parts. Then determine which transformation(s) map quadrilateral *KLMN* onto quadrilateral *FGHJ*.

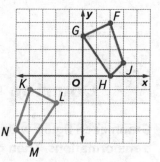

_____

_____

_____

_____

**14.** In the quilt design shown, $\triangle ABC \cong \triangle ADE$. What is the measure of $\angle BCA$?

_____

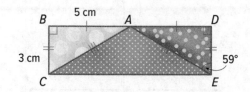

**15.** **MP** **Reason Abstractly** In the figure, $\triangle LZP \cong \triangle NZM$.

   **a.** On the figure, draw arc and tic marks to identify the corresponding parts.

   **b.** Find the value of *x*. _____

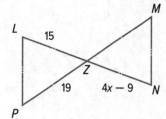

**16.** The triangles shown are congruent.

Complete the congruence statements to compare the corresponding parts.

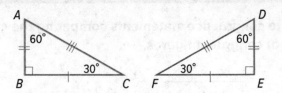

**a.** $\angle A \cong$ [　　　]　　**b.** $\angle B \cong$ [　　　]

**c.** $\angle C \cong$ [　　　]　　**d.** $\overline{AB} \cong$ [　　　]

**e.** $\overline{BC} \cong$ [　　　]　　**f.** $\overline{AC} \cong$ [　　　]

| $\angle A$ | $\angle D$ | $\overline{AB}$ | $\overline{DE}$ |
|---|---|---|---|
| $\angle B$ | $\angle E$ | $\overline{AC}$ | $\overline{DF}$ |
| $\angle C$ | $\angle F$ | $\overline{BC}$ | $\overline{EF}$ |

**17.** In the figure, $\triangle PQR \cong \triangle SQR$. Which of the following represent a congruence statement for the corresponding parts? Select all that apply.

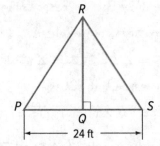

☐ $\angle RQP \cong \angle QSR$　　☐ $\overline{PQ} \cong \overline{RQ}$

☐ $\overline{RP} \cong \overline{RS}$　　☐ $\angle SRQ \cong \angle PRQ$

---

**CCSS** **Common Core Spiral Review**

Graph each figure with the given vertices and its image after the indicated transformations. Then give the coordinates of the final image. 8.G.3

**18.** $\triangle ABC$: $A(-4, 2)$, $B(-2, -3)$, $C(-4, -3)$; 90° counterclockwise rotation about $A$ followed by a translation of 4 units to the right

**19.** quadrilateral $RSTU$: $R(4, 3)$, $S(5, -1)$, $T(4, -3)$, $U(3, -1)$; reflection over the $x$-axis followed by a reflection over the $y$-axis

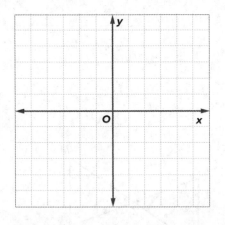

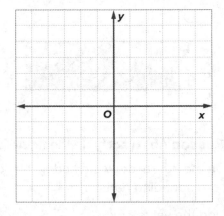

# Inquiry Lab
## Geometry Software

 **Inquiry** HOW can technology help you show the relationship between transformations and congruence?

 **Content Standards**
8.G.2

**MP Mathematical Practices**
1, 3, 5

While looking through a kaleidoscope at a flower, Evan noticed that some of the images were a result of a reflection followed by a rotation.

# Hands-On Activity

You can use The Geometer's Sketchpad® to perform transformations on a two-dimensional figure.

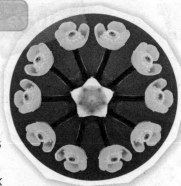

**Step 1** First, click on **Graph**. Go to **Grid Form** and click on **Square Grid.** Next, plot triangle *ABC* by clicking on **Graph** and then **Plot Points**. Enter the coordinates (3, 3), and then click **Plot**. Repeat the process for (1, 1) and (1, 4). Then click **Done**. Right click on point (3, 3) and click **Label Plotted Point**. Type **A** in the box provided and then click **OK**. Repeat this process for points *B*(1, 1) and *C*(1, 4).

**Step 2** Using the line segment tool, , click on one point and then another. A line segment will appear between those two points. Repeat until you have drawn a triangle like the one shown.

**Step 3** To reflect △*ABC* over the *x*-axis, use the selection arrow to click on the three points and three line segments that make up the triangle. Then double click on the *x*-axis. Use the pull-down menu under **Transform**. Click on **Reflect**. The program will automatically label the points of the image when you right click on a point and click **Label Plotted Point.** Draw the reflection on the grid shown.

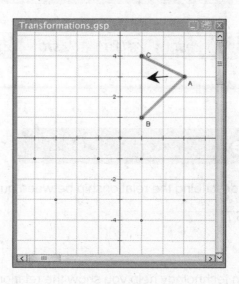

**Step 4** Rotate △*A'B'C'* by using the pull-down menu under **Transform**. Click **Rotate**. Type 270 when asked for the angle of rotation. Add labels to the points. Draw the rotation on the grid shown.

# Investigate

**Work with a partner. Once you have completed the transformations, draw the images that appear on your display.**

1. Graph quadrilateral *QRST* with vertices of *Q*(−5, −4), *R*(−4, −1), *S*(−2, −1), and *T*(−1, −3). Rotate the quadrilateral 180° about the origin and then reflect the image over the *y*-axis.

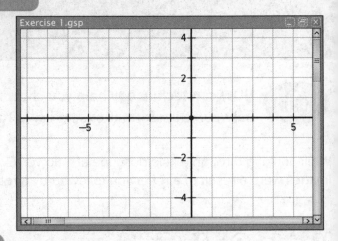

# Analyze and Reflect

**MP Use Math Tools Work with a partner.**

2. You can use The Geometer's Sketchpad® to measure the side lengths and angle measures of a figure. Use quadrilateral *QRST* from Exercise 1. Click on points *S* and *T*. On the MENU bar, click **Measure** and then **Distance**.

   What is *ST*? _____

3. You can also measure angles. Click on *S*, *T*, and *Q*. On the MENU bar, click

   **Measure** and then **Angle**. What is *m∠STQ*? _____

4. Measure the sides and angles of quadrilaterals *QRST* and *Q′R′S′T′*. Record the measures in the table.

| QR | Q″R″ | RS | R″S″ | ST | S″T″ | TQ | T″Q″ |
|----|------|----|------|----|------|----|------|
|    |      |    |      |    |      |    |      |

| ∠QRS | ∠Q″R″S″ | ∠RST | ∠R″S″T″ | ∠STQ | ∠S″T″Q″ | ∠TQR | ∠T″Q″ R″ |
|------|---------|------|---------|------|---------|------|----------|
|      |         |      |         |      |         |      |          |

# Create

5. Write a statement describing the relationship between quadrilateral *QRST* and

   quadrilateral *Q′R′S′T′*. _____

   _____

6. **Inquiry** HOW can technology help you show the relationship between transformations and congruence?

   _____

   _____

   _____

# MP Problem-Solving Investigation
# Draw a Diagram

## Case #1 Hammer Time

Christy wants to make shelves to store her game system and other electronics in her room. She will make brackets in the shape of right triangles to hold the shelves. Since it is a right triangle, one of the angles measure 90°.

*What is the relationship of the other two angles in a right triangle?*

**CCSS** Content Standards
8.G.5

**MP** Mathematical Practices
1, 3, 4

## Understand *What are the facts?*

The bracket is in the shape of a right triangle, so one of the angles measures 90°.

## Plan *What is your strategy to solve this problem?*

Draw several right triangles, measure each angle, and look for a pattern.

## Solve *How can you apply the strategy?*

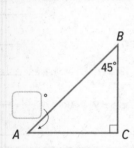
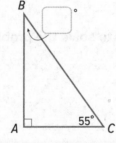
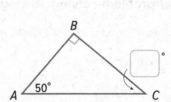

It appears that the sum of the measures of the acute angles of a right

triangle is ☐. So, the acute angles are _____.

## Check *Does the answer make sense?*

You can try several more examples to see whether your conjecture appears to be true. But at this point, it is just a conjecture, not an actual proof.

## Analyze the Strategy   Tutor

**MP** **Justify Conclusions** Inductive reasoning is the process of making a conjecture after observing several examples. Did Christy use inductive

reasoning? Explain. _____

_____

## Case #2  Bike-A-Thon

Jacob is participating in a biking fundraiser to the lake. After 45 miles, he is $\frac{5}{6}$ of the way there.

How many more miles does he need to travel to reach the lake?

### 1 Understand

**Read the problem. What are you being asked to find?**

I need to find _____.

**Underline key words and values. What information do you know?**

Jacob has biked _____ of the way to the lake. This is equal

to _____.

### 2 Plan

**Choose a problem-solving strategy.**

I will use the _____ strategy.

### 3 Solve

**Use your problem-solving strategy to solve the problem.**

Draw a line that represents the distance to the lake. Divide the line

into 6 equal parts.

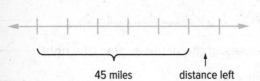

45 miles        distance left

_____ of the 6 parts = 45 so

each part is _____ miles.

_____ + _____ + _____ +

_____ + _____ = 45

The distance to the lake is

45 + _____ = _____ miles.

So, Jacob has _____ left to ride.

### 4 Check

**Use information from the problem to check your answer.**

_____

**Work with a small group to solve the following cases.**
**Show your work on a separate piece of paper.**

Collaborate

## Case #3 Dance

Ms. Samson's dance class is evenly shaped in a circle.

If the sixth person is directly opposite the sixteenth person, how many people are in the circle?

_____

## Case #4 Stadium Seating

A section of a baseball stadium is set up so that each row has the same number of seats. Kyleigh is seated in the seventh row from the back and the eighth row from the front of this section. Her seat is the fourth row from the right and the seventh from the left.

How many seats are in this section of the stadium?

_____

## Case #5 Scrapbooks

A scrapbook page measures 12 inches long by 12 inches wide.

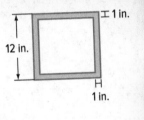

How many 3-inch by 5-inch horizontal photographs can be placed on the page if $\frac{1}{2}$ inch is placed between each photo and at least 1 inch is left as a margin on all four sides?

_____

## Case #6 Geometry

The sides of a right triangle are in the ratio 3: 4: 5. The perimeter of the triangle is 84 feet.

What is the area of the triangle?

_____

Use any strategy!

# Mid-Chapter Check

## Vocabulary Check

1. What transformations can be used to show two figures are congruent? (Lesson 1)

_____

2. List two attributes of two congruent polygons. (Lesson 2)

_____

_____

## Skills Check and Problem Solving

**Determine if the two figures are congruent by using transformations. Explain your reasoning.** (Lesson 1)

3.

4.

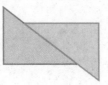

5.

_____    _____    _____

_____    _____    _____

_____    _____    _____

6. Jordan is creating the logo shown using a pentagon and five congruent triangles. Triangle *WAX* is congruent to triangle *YBZ*. Describe the transformations that map △*WAX* onto △*YBZ*. If *WX* measures 5 inches, what is the length of *YZ*? (Lesson 2)

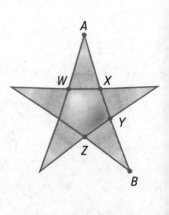

_____

7. (MP) **Persevere with Problems** Trapezoid *MNOP* is congruent to trapezoid *QROP*. Which transformation maps *MNOP* onto *QROP*? (Lesson 2)

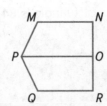

_____

# Inquiry Lab
## Similar Triangles

**Inquiry** HOW are two triangles related if they have the same shape but different sizes?

**CCS** Content Standards
Preparation for 8.G.4

**MP** Mathematical Practices
1, 3

While flying in an airplane, Ariel looked out the window and saw roads and a field like the one shown. She wondered if there was a relationship between the two triangles she saw.

# Hands-On Activity

To determine if there is a relationship between the two triangles, use the diagram shown.

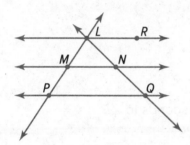

$\overleftrightarrow{LR} \parallel \overleftrightarrow{MN} \parallel \overleftrightarrow{PQ}$
$\overleftrightarrow{LP}$ and $\overleftrightarrow{LQ}$ are transversals.

**Step 1** Measure and record the lengths of the line segments in millimeters and angles in degrees in the table.

| △LPQ | | | △LMN | | |
|---|---|---|---|---|---|
| LP = | m∠L = ° | LM = | m∠L = ° | | |
| LQ = | m∠P = ° | LN = | m∠M = ° | | |
| PQ = | m∠Q = ° | MN = | m∠N = ° | | |

What do you notice about the measure of the corresponding angles of the triangles? _____

**Step 2** Express the lengths of the corresponding sides of the triangles as ratios.

$\dfrac{LP}{LM} =$ _____   $\dfrac{LQ}{LN} =$ _____   $\dfrac{PQ}{MN} =$ _____

What do you notice about the ratios of the corresponding sides of the triangles? _____

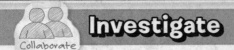

**Work with a partner.**

1.  **Model with Mathematics** Triangle *ABC* is a right triangle with
   $m\angle A = 53°$. On the grid, draw and label a different right triangle,
   *XYZ*, using the given angle *X*, which also measures 53°.

Show your work.

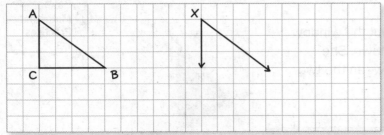

What do you notice about the shape of the triangles? _____

_____

 **Analyze and Reflect**

**For Exercises 2–4, refer to the triangles in Exercise 1.**

2. What is the measure of $\angle B$? the measure of the angle that corresponds to $\angle B$

   in $\triangle XYZ$? _____

3. Express the lengths of the corresponding sides of the triangles as ratios.

   $$\frac{AC}{\boxed{\phantom{0}}} = \frac{\boxed{\phantom{0}}}{\boxed{\phantom{0}}}$$  $$\frac{CB}{\boxed{\phantom{0}}} = \frac{\boxed{\phantom{0}}}{\boxed{\phantom{0}}}$$  $$\frac{AB}{\boxed{\phantom{0}}} = \frac{\boxed{\phantom{0}}}{\boxed{\phantom{0}}}$$

4. What do you notice about the ratios? _____

**Create**

On Your Own

5. **MP** **Reason Inductively** The two triangles in the Activity and in Exercise 1 are
   called *similar triangles*. Based on your discoveries, make a conjecture about
   the properties of similar triangles.

   _____

   _____

6. **Inquiry** HOW are two triangles related if they have the same shape but
   different sizes?

   _____

# Similarity and Transformations

## Vocabulary Start-Up

Recall that a dilation changes the size of a figure by a scale factor, but does not change the shape of the figure. Since the size is changed, the image and the preimage are not congruent.

**Complete the graphic organizer. Consider each word on the Rating Scale and place a check ✔ in the appropriate column next to the word. If you do not know the meaning of a word, find the meaning in the glossary or on the Internet.**

| | Rating Scale | | | |
|---|---|---|---|---|
| **Word** | **Know it well** | **Have seen or heard it** | **No clue** | **What it means** |
| dilation | | | | |
| scale factor | | | | |
| similar figures | | | | |

### Essential Question

HOW can you determine congruence and similarity?

### Vocabulary

similar

### Common Core State Standards

**Content Standards**
8.G.4

(MP) **Mathematical Practices**
1, 3, 4, 7

## Real-World Link

A *fractal* is a geometric image that can be divided into parts that are smaller copies of the whole. The photo at the right is an example of a fractal.

1. (Circle) two different size parts of the figure that are smaller copies of the whole.

**Which (MP) Mathematical Practices did you use?**
**Shade the circle(s) that applies.**

① Persevere with Problems
② Reason Abstractly
③ Construct an Argument
④ Model with Mathematics

⑤ Use Math Tools
⑥ Attend to Precision
⑦ Make Use of Structure
⑧ Use Repeated Reasoning

# Identify Similarity

Two figures are **similar** if the second can be obtained from the first by a sequence of transformations and dilations.

## Examples

Tutor

**1.** **Determine if the two triangles are similar by using transformations.**

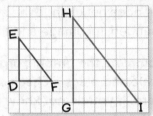

Since the orientation of the figures is the same, one of the transformations might be a translation.

**Step 1** Translate △DEF down 2 units and 5 units to the right so D maps onto G.

**Step 2** Write ratios comparing the lengths of each side.

$\frac{HG}{ED} = \frac{8}{4}$ or $\frac{2}{1}$    $\frac{GI}{DF} = \frac{6}{3}$ or $\frac{2}{1}$    $\frac{IH}{FE} = \frac{10}{5}$ or $\frac{2}{1}$

Since the ratios are equal, △HGI is the dilated image of △EDF. So, the two triangles are similar because a translation and a dilation maps △EDF onto △HGI.

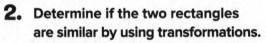

**2.** **Determine if the two rectangles are similar by using transformations.**

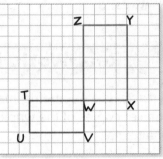

The orientation of the figures is the same, so one of the transformations might be a rotation.

**Step 1** Rotate rectangle VWTU 90° clockwise about W so that it is oriented the same way as rectangle WXYZ.

**Step 2** Write ratios comparing the lengths of each side.

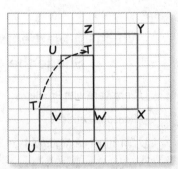

$\frac{WT}{XY} = \frac{5}{7}$    $\frac{TU}{YZ} = \frac{3}{4}$

$\frac{UV}{ZW} = \frac{5}{7}$    $\frac{VW}{WX} = \frac{3}{4}$

The ratios are not equal. So, the two rectangles are not similar since a dilation did not occur.

**Got it?** Do these problems to find out.

a.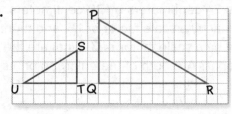

b.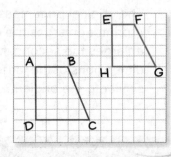

a. _____

b. _____

# Use the Scale Factor

Similar figures have the same shape, but may have different sizes. The sizes of the two figures are related to the scale factor of the dilation.

| If the scale factor of the dilation is ... | then the dilated figure is ... |
|---|---|
| between 0 and 1 | smaller than the original |
| equal to 1 | the same size as the original |
| greater than 1 | larger than the original |

## Example

Tutor

**3.** Ken enlarges the photo shown by a scale factor of 2 for his webpage. He then enlarges the webpage photo by a scale factor of 1.5 to print. If the original photo is 2 inches by 3 inches, what are the dimensions of the print? Are the enlarged photos similar to the original?

Multiply each dimension of the original photo by 2 to find the dimensions of the webpage photo.

2 in. × 2 = 4 in.          3 in. × 2 = 6 in.

So, the webpage photo will be 4 inches by 6 inches. Multiply the dimensions of that photo by 1.5 to find the dimensions of the print.

4 in. × 1.5 = 6 in.          6 in. × 1.5 = 9 in.

The printed photo will be 6 inches by 9 inches. All three photos are similar since each enlargement was the result of a dilation.

**STOP and Reflect**

List below at least two topics in mathematics that use a scale factor.

c. _____

c. An art show offers different size prints of the same painting. The original print measures 24 centimeters by 30 centimeters. A printer enlarges the original by a scale factor of 1.5, and then enlarges the second image by a scale factor of 3. What are the dimensions of the largest print? Are both of the enlarged prints similar to the original?

## Guided Practice

**Determine if the two figures are similar by using transformations. Explain your reasoning.** (Examples 1 and 2)

1.

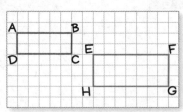

_____

_____

_____

2.

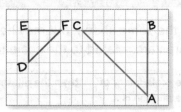

_____

_____

_____

3. A T-shirt iron-on measures 2 inches by 1 inch. It is enlarged by a scale factor of 3 for the back of the shirt. The second iron-on is enlarged by a scale factor of 2 for the front of the shirt. What are the dimensions of the largest iron-on? Are both of the enlarged iron-ons similar to the original? (Example 3) _____

4.  **Building on the Essential Question** What is the difference between using transformations to create similar figures versus using transformations to create congruent figures?

_____

_____

_____

_____

# Independent Practice

Go online for Step-by-Step Solutions

**Determine if the two figures are similar by using transformations.**
**Explain your reasoning.** (Examples 1 and 2)

**1**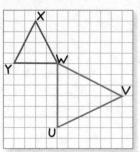

**2.**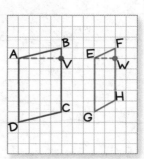

_____
_____
_____

_____
_____
_____

**3** Felisa is creating a scrapbook of her family. A photo of her grandmother measures
3 inches by 5 inches. She enlarges it by a scale factor of 1.5 to place in the
scrapbook. Then she enlarges the second photo by a scale factor of 1.5 to place on
the cover of the scrapbook. What are the dimensions of the photo for the cover of

the scrapbook? Are all of the photos similar? (Example 3) _____

**MP Persevere with Problems** Each preimage and image are similar.
Describe a sequence of transformations that maps the preimage onto
the image.

**4.**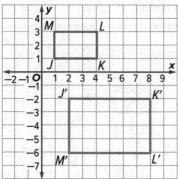

**5.**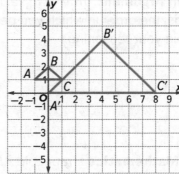

_____
_____
_____

_____
_____
_____

6. **MP Identify Structure** Use the graphic organizer to compare and contrast similar and congruent figures.

| | Similar Figures | Congruent Figures |
|---|---|---|
| Side Measures | | |
| Angle Measures | | |
| Transformations Used | | |

## H.O.T. Problems Higher Order Thinking

7. **MP Persevere with Problems** Using at least one dilation, describe a series of transformations where the image is congruent to the preimage.

_____

_____

_____

8. **MP Model with Mathematics** The image of $\triangle DEF$ after two transformations has vertices at $D'(3, 3)$, $E'(6, 3)$ and $F'(3, -6)$. If the two triangles are similar, determine what two transformations map $\triangle DEF$ onto $\triangle D'E'F'$.

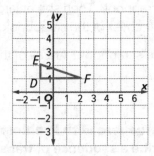

_____

_____

_____

9. **MP Construct an Argument** _True_ or _false_. If a dilation is in a composition of transformations, the order in which you perform the composition does not matter. Explain your reasoning.

_____

_____

_____

10. **MP Model with Mathematics** Trapezoid _ABCD_ is shown at the right. Perform a series of transformations on the trapezoid and draw the image on the coordinate plane. List the transformations used below.

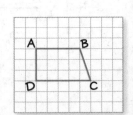

_____

_____

# Extra Practice

**Determine if the two figures are similar by using transformations. Explain your reasoning.**

11.

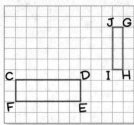

12.

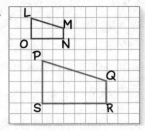

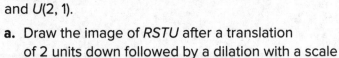

no; The ratios of the side lengths are

not equal.

Find the ratios of the side lengths.

$\frac{CD}{GH} = \frac{6}{4}$ and $\frac{DE}{JG} = \frac{2}{1}$; $\frac{6}{4} \neq \frac{2}{1}$, so the two

figures are not similar.

13. Shannon is making three different sizes of blankets from the same material. The first measures 2.5 feet by 2 feet. She wants to enlarge it by a scale factor of 2 to make the second blanket. Then she will enlarge the second one by a scale factor of 1.5 to make the third blanket. What are the dimensions of the third blanket? Are all of the blankets similar?

_____

14. **MP Model with Mathematics** In the figure shown, trapezoid *RSTU* has vertices *R*(1, 3), *S*(4, 3), *T*(3, 1), and *U*(2, 1).

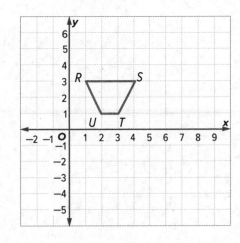

a. Draw the image of *RSTU* after a translation of 2 units down followed by a dilation with a scale factor of 2. Label the vertices *ABCD*.

b. Draw the image of *RSTU* after a dilation with a scale factor of 2, followed by a translation of 2 units down. Label the vertices *EFGH*.

c. Which figures are similar? Which figures are congruent?

_____

_____

d. Are *ABCD* and *EFGH* in the same location? If they are not, what transformation would map *ABCD* onto *EFGH*?

_____

_____

**15.** Triangle *DEF* is the image of triangle *ABC* after a sequence of transformations.

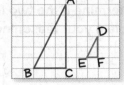

Determine if each statement is true or false.

**a.** △*ABC* was dilated by a scale factor of 3 to create △*DEF*.  ☐ True  ☐ False

**b.** The transformation represents a reduction.  ☐ True  ☐ False

**c.** The ratios $\frac{AB}{DF}$ and $\frac{AC}{DE}$ are equal.  ☐ True  ☐ False

**16.** Which sequences of transformations would result in similar figures that are enlargements or reductions? Select all that apply.

☐ translation, dilation, rotation, reflection

☐ reflection, translation, rotation

☐ translation, reflection, rotation, reflection

☐ rotation, translation, dilation

CCSS **Common Core Spiral Review**

**Find the coordinates of the vertices of each figure after a dilation with the given scale factor *k*. Then graph the original image and the dilation.** 8.G.3

**17.** *M*(0, 0), *N*(−1, 1), *O*(2, 3); $k = 2$

**18.** *A*(−3, 3), *B*(3, 3), *C*(3, −3); $k = \frac{2}{3}$

**19.** *G*(4, 4), *H*(2, −4), *I*(−4, −4), *J*(0, 2); $k = \frac{1}{2}$

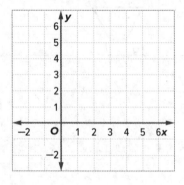

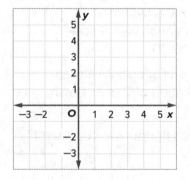

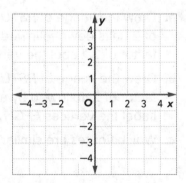

# Properties of Similar Polygons

## Real-World Link

**Photos** Elsa is printing pictures at a photo kiosk in the store. She can choose between 4 × 6 prints or 5 × 7 prints. Are the side lengths of the two prints proportional? Explain. _____

**Collaborate** Follow the steps to discover how the triangles are related.

1. Using a centimeter ruler, measure the sides of the two triangles. Then, use a protractor to measure the angles. Write the results in the table.

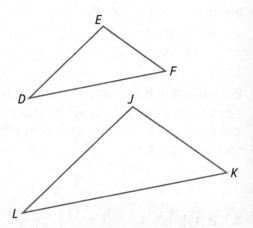

### Essential Question

HOW can you determine congruence and similarity?

**Vocab**

### Vocabulary

similar polygons
scale factor

**Math Symbols**

~ is similar to

### Common Core State Standards

**Content Standards**
8.G.4

**MP** **Mathematical Practices**
1, 2, 3, 4

| Figure | Side Length (cm) | | | Angle Measure (°) | | |
|---|---|---|---|---|---|---|
| △EFD | DE | EF | FD | ∠D | ∠E | ∠F |
| | | | | | | |
| △LJK | LJ | JK | KL | ∠L | ∠J | ∠K |
| | | | | | | |

2. Are the side lengths proportional? Explain.

_____

3. What do you notice about the angles of the two triangles?

_____

### Which MP **Mathematical Practices** did you use?
### Shade the circle(s) that applies.

① Persevere with Problems     ⑤ Use Math Tools

② Reason Abstractly     ⑥ Attend to Precision

③ Construct an Argument     ⑦ Make Use of Structure

④ Model with Mathematics     ⑧ Use Repeated Reasoning

# Similar Polygons

**Work Zone**

**Words**   If two polygons are similar, then
- their corresponding angles are congruent and
- the measures of their corresponding sides are proportional.

**Model**

$\triangle ABC \sim \triangle XYZ$

**Symbols**   $\angle A \cong \angle X$, $\angle B \cong \angle Y$, $\angle C \cong \angle Z$, and $\dfrac{AB}{XY} = \dfrac{BC}{YZ} = \dfrac{AC}{XZ}$

Polygons that have the same shape are called **similar polygons**. In the Key Concept box, triangle *ABC* is similar to triangle *XYZ*. This is written as △*ABC* ~ △*XYZ*. The parts of similar figures that "match" are called corresponding parts.

## Example

1. **Determine whether rectangle *HJKL* is similar to rectangle *MNPQ*. Explain.**

   First, check to see if corresponding angles are congruent.

   Since the two polygons are rectangles, all of their angles are right angles. Therefore, all corresponding angles are congruent.

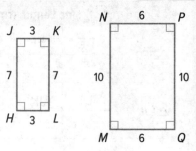

   Next, check to see if corresponding sides are proportional.

   $$\frac{HJ}{MN} = \frac{7}{10} \qquad \frac{JK}{NP} = \frac{3}{6} \text{ or } \frac{1}{2} \qquad \frac{KL}{PQ} = \frac{7}{10} \qquad \frac{LH}{QM} = \frac{3}{6} \text{ or } \frac{1}{2}$$

   Since $\frac{7}{10}$ and $\frac{1}{2}$ are not equivalent, the rectangles are *not* similar.

**Common Error**

Do not assume that two rectangles are similar just because their corresponding angles are congruent. Their corresponding sides must also be proportional.

*Show your work.*

**Got it?**  Do this problem to find out.

a. _____

a. Determine whether △*ABC* is similar to △*XYZ*. Explain.

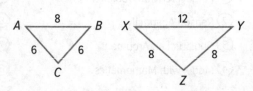

# Find Missing Measures

**Scale factor** is the ratio of the lengths of two corresponding sides of two similar polygons. You can use the scale factor of similar figures to find missing measures.

# Example

Tutor

2. Quadrilateral *WXYZ* is similar to quadrilateral *ABCD*.

 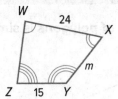

a. **Describe the transformations that map *WXYZ* onto *ABCD*.**

Since the figures are similar, they are not the same size. Choose two corresponding sides and determine what transformations will map one onto the other. A translation followed by a dilation will map $\overline{AB}$ onto $\overline{WX}$.

b. **Find the missing measure.**

**Method 1**

Find the scale factor from quadrilateral *ABCD* to quadrilateral *WXYZ*.

$$\text{scale factor: } \frac{YZ}{CD} = \frac{15}{10} \text{ or } \frac{3}{2}$$

So, a length on polygon *WXYZ* is $\frac{3}{2}$ times as long as the corresponding length on polygon *ABCD*. Let *m* represent the measure of $\overline{XY}$.

$m = \frac{3}{2}(12)$    Write the equation.

$m = 18$    Multiply.

**Method 2**

Set up a proportion to find the missing measure.

$\dfrac{XY}{BC} = \dfrac{YZ}{CD}$    Write the proportion.

$\dfrac{m}{12} = \dfrac{15}{10}$    $XY = m, BC = 12, YZ = 15, CD = 10$

$m \cdot 10 = 12 \cdot 15$    Find the cross products.

$10m = 180$    Simplify.

$m = 18$    Division Property of Equality

b. _____

c. _____

**Got it?** Do these problems to find out.

**Find each missing measure.**

b. *WZ*

c. *AB*

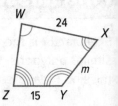

## Guided Practice

**Determine whether each pair of polygons is similar. Explain.** (Example 1)

1.

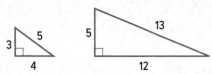

_____

_____

2.

_____

_____

3. The two triangles are similar. (Example 2)

   a. Determine the transformations that map one figure onto the other.

   _____

   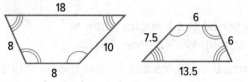

   b. Find the missing side measures. _____

4. The two triangles are similar. (Example 2)

   a. Determine the transformations that map one figure onto the other.

   _____

   b. Find the missing side measure. _____

**Rate Yourself!**

Are you ready to move on? Shade the section that applies.

I have a few questions.

I'm ready to move on.

I have a lot of questions.

5. **Building on the Essential Question** How does the scale factor of a dilation relate to the ratio of two of the corresponding sides of the preimage and the image?

_____

_____

For more help, go online to access a Personal Tutor.

Tutor

**FOLDABLES** Time to update your Foldable!

# Independent Practice

Go online for Step-by-Step Solutions  eHelp

**Determine whether each pair of polygons is similar. Explain.** (Example 1)

**1.**

3
4
7
8

_____

_____

_____

**2.**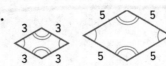

3    3
3    3

5    5
5    5

_____

_____

_____

**Each pair of polygons is similar. Determine the transformations that map one figure onto the other. Then find the missing side measures.** (Example 2)

**3.**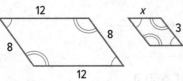

12
8        8
12

x
3

_____

_____

**4.**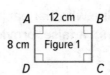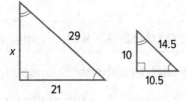

29
x
21

14.5
10
10.5

_____

_____

**5.** (MP) **Persevere with Problems** The figures at the right are similar.

**a.** Find the area of both figures.

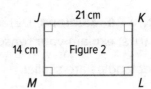

A   12 cm   B
8 cm   Figure 1
D           C

J   21 cm   K
14 cm   Figure 2
M           L

_____

**b.** Compare the scale factor of the side lengths and the ratio of the areas.

_____

_____

_____

**6.** **STEM** The scale factor from the model of a human inner ear to the actual ear is 55:2. If one of the bones of the model is 8.25 centimeters

long, how long is the actual bone in a human ear? _____

7. **(MP) Model with Mathematics** Refer to the graphic novel frame below. The brochure says that the rope is 500 feet long. Use the properties of similar triangles to find the parasailer's height above the water. _____

Hmmm. I wonder how far we'll be above the water?

8. **(MP) Persevere with Problems** Suppose two rectangles are similar with a scale factor of 2. What is the ratio of their areas? Explain. _____

_____

_____

**(MP) Justify Conclusions** Determine whether each statement is *true* or *false*. If true, explain your reasoning. If false, provide a counterexample.

9. All rectangles are similar.

_____

_____

_____

_____

10. All squares are similar.

_____

_____

_____

_____

11. **(MP) Model with Mathematics** Draw two similar polygons in the space provided. Include the measures of the sides on your drawing, and identify the scale factor. _____

# Extra Practice

**Determine whether each pair of polygons is similar. Explain.**

**12.**

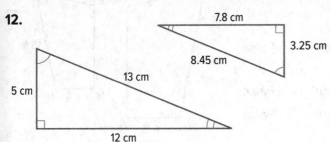

7.8 cm

3.25 cm

8.45 cm

13 cm

5 cm

12 cm

As indicated by the arc marks, corresponding angles are congruent. Check to see if the corresponding sides are proportional.

$$\frac{3.25}{5} = \frac{8.45}{13} = \frac{7.8}{12}$$

The sides are proportional so the triangles are similar.

**13.**

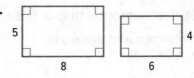

5

8

4

6

**14.** The two figures are similar. Determine the transformations that map one figure onto the other. Then find the missing side length.

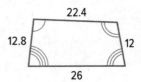

22.4

12.8

12

26

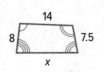

14

8

7.5

x

**15.** **MP** **Model with Mathematics**

Mrs. Henderson wants to build a fence around the rectangular garden in her backyard. In the scale drawing, the perimeter of the garden is 14 inches. If the actual length of $\overline{AB}$ is 20 feet, how many feet of fencing will she need?

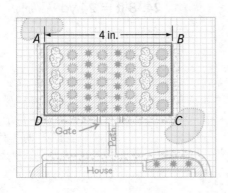

A  |———— 4 in. ————|  B

D

Gate

C

Path

House

**16.** Isaiah is making a mosaic using different pieces of tile. The tiles shown at the right are similar. If the perimeter of the larger tile is 23 centimeters, what is the perimeter of the smaller tile?

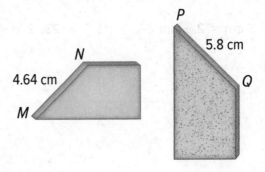

P

5.8 cm

N

Q

4.64 cm

M

**17.** Quadrilateral *FGHJ* was transformed to create similar quadrilateral *LMNO*.

Determine if each statement is true or false.

**a.** *FGHJ* was reflected and dilated to create *LMNO*.  ☐ True  ☐ False

**b.** The scale factor of the dilation is $\frac{3}{4}$.  ☐ True  ☐ False

**c.** The value of *x* is 16.  ☐ True  ☐ False

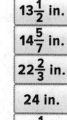

**18.** Triangle *FGH* is similar to triangle *RST*. Select the correct values to label the missing side lengths of triangle *RST*.

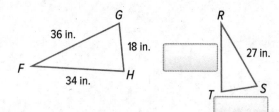

| |
|---|
| $13\frac{1}{2}$ in. |
| $14\frac{5}{7}$ in. |
| $22\frac{2}{3}$ in. |
| 24 in. |
| $25\frac{1}{2}$ in. |

**Find the scale factor for each scale drawing.** 7.G.1

**19.** 6 in. = 12 ft _____

**20.** 20 cm = 10 m _____

**21.** 18 in. = 3 ft _____

**22.** 8 cm = 2.5 mm _____

**23.** 2 in. = 0.25 mi _____

**24.** 8 ft = 24 yd _____

# Similar Triangles and Indirect Measurement

## Vocabulary Start-Up

**Indirect measurement** allows you to use properties of similar polygons to find distances or lengths that are difficult to measure directly.

**Complete the graphic organizer. List three real-world examples in the Venn diagram for each method of measurement.**

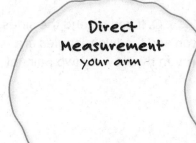

Direct Measurement
your arm

Indirect Measurement
Statue of Liberty's arm

Write the name of an object that could be measured by either

method. _____

 ## Real-World Link

**Shadows** Legend says that Thales, the first Greek mathematician, was the first to determine the height of the pyramids by examining the shadows made by the Sun.

1. What appears to be true about the corresponding

   angles in the two triangles? _____

2. If the corresponding sides are proportional, what could you

   conclude about the triangles? _____

**Which MP Mathematical Practices did you use?**
**Shade the circle(s) that applies.**

① Persevere with Problems      ⑤ Use Math Tools

② Reason Abstractly            ⑥ Attend to Precision

③ Construct an Argument        ⑦ Make Use of Structure

④ Model with Mathematics       ⑧ Use Repeated Reasoning

 **Essential Question**

HOW can you determine congruence and similarity?

 **Vocabulary**

indirect measurement

 **Common Core State Standards**

**Content Standards**
8.G.5

 **Mathematical Practices**
1, 3, 4, 7

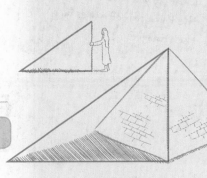

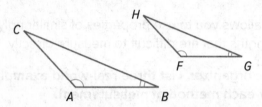

## Key Concept ⟩ Angle-Angle (AA) Similarity

**Words**      If two angles of one triangle are congruent to two angles of another triangle, then the triangles are similar.

**Symbols**    If ∠A ≅ ∠F and ∠B ≅ ∠G, then △ABC ~ △FGH.

**Model**

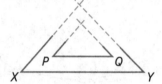

*Work Zone*

In the figure below, ∠X ≅ ∠P and ∠Y ≅ ∠Q. If you extend the sides of each figure to form a triangle, you can see the two triangles are similar. So, triangle similarity can be proven by showing two pairs of corresponding angles are congruent.

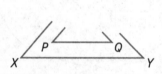

**STOP and Reflect**

What do you know about the third pair of angles in the triangle?

**Example** 🗨 Tutor

1. **Determine whether the triangles are similar. If so, write a similarity statement.**

   Angle A and ∠E have the same measure, so they are congruent. Since 180 − 62 − 48 = 70, ∠G measures 70°. Two angles of △EFG are congruent to two angles of △ABC, so △ABC ~ △EFG.

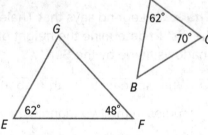

**Got it?** Do this problem to find out.

a. _____

a.

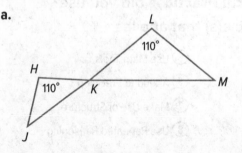

Show your work.

# Use Indirect Measurement

One type of indirect measurement is *shadow reckoning*. Two objects and their shadows form two sides of right triangles. In shadow problems, you can assume that the angles formed by the Sun's rays with two objects at the same location are congruent. Since two pairs of corresponding angles are congruent, the two right triangles are similar. You can also use similar triangles that do not involve shadows to find missing measures.

## Examples

Watch  Tutor

**2.** A fire hydrant 2.5 feet high casts a 5-foot shadow. How tall is a street light that casts a 26-foot shadow at the same time? Let *h* represent the height of the street light.

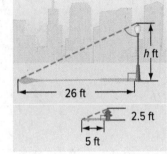

**Shadow**       **Height**

hydrant ⟶   $\dfrac{5}{26} = \dfrac{2.5}{h}$   ⟵ hydrant
street light ⟶           ⟵ street light

$5h = 26 \cdot 2.5$     Find the cross products.

$5h = 65$     Multiply.

$\dfrac{5h}{5} = \dfrac{65}{5}$     Divide each side by 5.

$h = 13$

The street light is 13 feet tall.

**3.** In the figure at the right, triangle *DBA* is similar to triangle *ECA*. Ramon wants to know the distance across the lake.

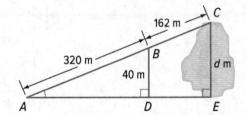

$\dfrac{AB}{AC} = \dfrac{BD}{CE}$     $\overline{AB}$ corresponds to $\overline{AC}$ and $\overline{BD}$ corresponds to $\overline{CE}$.

$\dfrac{320}{482} = \dfrac{40}{d}$     Replace *AB* with 320, *AC* with 482, and *BD* with 40.

$320d = 482 \cdot 40$     Find the cross products.

$\dfrac{320d}{320} = \dfrac{19,280}{320}$     Multiply. Then divide each side by 320.

$d = 60.25$

The distance across the lake is 60.25 meters.

**Got it?** Do this problem to find out.

b. _____

**b.** At the same time a 2-meter street sign casts a 3-meter shadow, a nearby telephone pole casts a 12.3-meter shadow. How tall is the telephone pole?

## Guided Practice

Check ✓

**Determine whether the triangles are similar. If so, write a similarity statement.** (Example 1)

**1.**

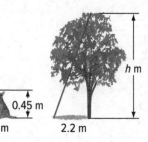

Show your work.

_____

_____

**2.**

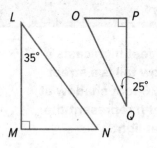

_____

_____

**3.** How tall is the tree? (Example 2) _____

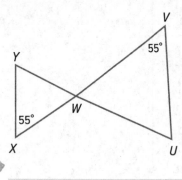

**4.** Find the distance from the house to the street light. (Example 3) _____

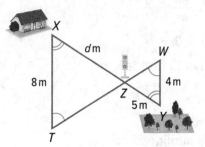

**5.** 🇪 **Building on the Essential Question** How do similar triangles make it easier to measure very tall objects?

_____

_____

_____

_____

_____

**Rate Yourself!**

Are you ready to move on? Shade the section that applies.

YES ? NO

For more help, go online to access a Personal Tutor.

Tutor

# Independent Practice

Go online for Step-by-Step Solutions

**Determine whether the triangles are similar. If so, write a similarity statement.** (Example 1)

**1.**

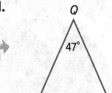

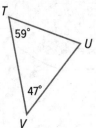

Q
47°
R   68°   S
T
59°
U
47°
V

_____

_____

**2.**

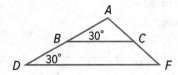

A
B   30°   C
D   30°   F

_____

_____

**3.** How tall is the building? (Example 2)

_____

h ft
50 ft
50 ft   12.5 ft

50 ft

**4.** How tall is the taller flagpole? (Example 2)

_____

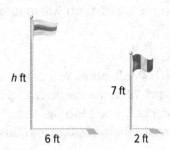

h ft
7 ft
6 ft   2 ft

**5** How far is it from the log ride to the pirate ship? (Example 3) _____

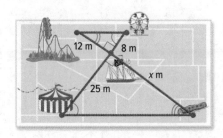

12 m   8 m
x m
25 m

**6.** Find the height of the brace. (Example 3)

_____

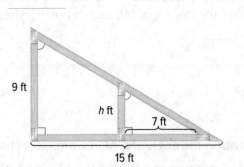

9 ft
h ft
7 ft
15 ft

**7** **MP** **Reason Abstractly** The Giant Wheel at Cedar Point in Ohio is one of the tallest Ferris wheels in the country at 136 feet tall. If the Giant Wheel casts a 34-foot shadow, write and solve a proportion to find the height of a nearby man who casts a $1\frac{1}{2}$-foot shadow.

_____

_____

8. MP **Find the Error** Sara is finding the height of the lighthouse shown in the diagram. Find her mistake and correct it.

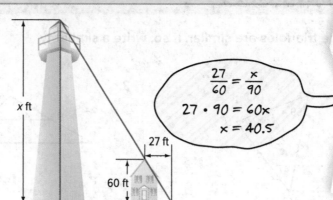

$$\frac{27}{60} = \frac{x}{90}$$

$$27 \cdot 90 = 60x$$

$$x = 40.5$$

_____

_____

_____

_____

_____

9. MP **Model with Mathematics** On a separate sheet of paper, draw two different triangles so that each one contains both of the angles shown. Then verify that they are similar by determining which transformation will map one onto the other.

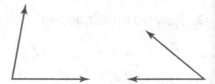

10. MP **Persevere with Problems** You cut a circular hole $\frac{1}{4}$-inch in diameter in a piece of cardboard. With the cardboard 30 inches from your face, the Moon fits exactly into the hole. The Moon is about 240,000 miles from Earth. Is the Moon's diameter more than 1,500 miles? Justify your reasoning.

_____

_____

11. MP **Identify Structure** What measures must be known in order to calculate the height of tall objects using shadow reckoning?

_____

_____

_____

12. MP **Reason Inductively** Mila wants to estimate the height of a statue in a local park. Mila's height and both shadow lengths are shown in the diagram. Is an estimate of 15 feet reasonable for the statue's height? Explain your reasoning.

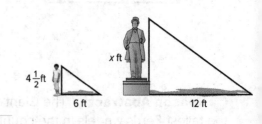

_____

_____

_____

_____

# Extra Practice

**13.** What is the height of the tree? _90 ft_

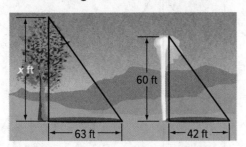

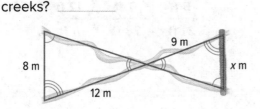

mework Help → The triangles are similar. Write and solve a proportion.

$$\frac{63}{42} = \frac{x}{60}$$

$$63 \cdot 60 = 42x$$

$$90 = x$$

**14.** Find the distance across the river. _____

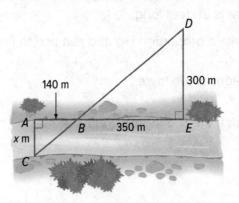

**15.** About how long is the log that goes across the creeks? _____

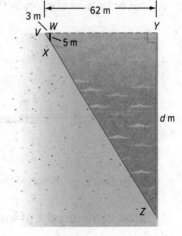

**16.** How deep is the water 62 meters from the shore?

_____

**17.** In the diagram shown at the right, △ABC ~ △EDC.

a. Write a proportion that could be used to solve for the height *h* of the flag pole. _____

b. What information would you need to know in order to solve this proportion?

_____

_____

_____

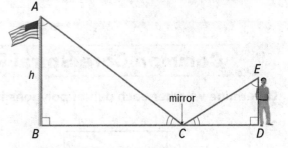

**18.** ⓂⓅ **Model with Mathematics** A 78-inch-tall man casts a shadow that is 54 inches long. At the same time, a nearby building casts a 48-foot-long shadow. Write and solve a proportion to find the height of the building.

_____

**19.** Horatio is 6 feet tall and casts a shadow 3 feet long. At the same time, a nearby tower casts a shadow that is 25 feet long.

Write a proportion Horatio can use to find the height of the tower.

Using the proportion, the tower is [        ] feet tall.

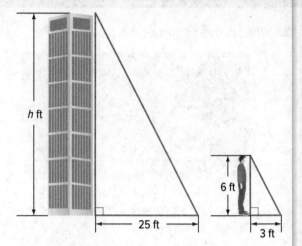

*h* ft

25 ft

6 ft

3 ft

**20.** Lenno is 5 feet tall and is using similar triangles and a mirror to find the height of a telephone pole. The horizontal distance between Lenno and the telephone pole is 28 feet. He places the mirror on the ground 7 feet from himself so that he can see the top of the pole in the mirror's reflection as shown in the figure below.

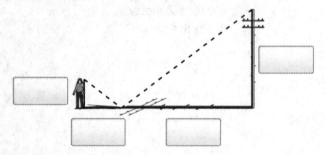

| 5 ft | 7 ft | 12 ft | 14 ft |
|------|------|-------|-------|
| 21 ft | 28 ft | *h* ft | |

Select values to label the diagram with the correct dimensions.

What is the height of the telephone pole? [        ]

**Determine whether each pair of polygons is similar. Explain.** 8.G.4

**21.**

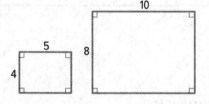

10

5

8

4

**22.**

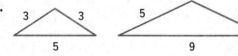

3      3

5

5      5

9

_____

_____

# Slope and Similar Triangles

## Real-World Link

**Physics** In an experiment using a coiled spring toy, Zoe and Jack determined they needed to raise one side of a 5-foot board 3 feet for the toy to move.

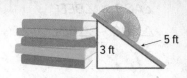

1. Find the slope of the board. (*Hint:* Use the Pythagorean Theorem to find how far the end of the board is from the books.) _____

Collaborate

**Work with a partner. Use the graph to discover how slope triangles are related.**

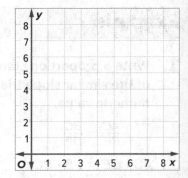

1. Draw the triangle formed by $A(0, 2)$, $B(0, 4)$, and $C(3, 4)$. What kind of triangle did you draw?

_____

2. Draw the triangle formed by $D(6, 6)$, $F(6, 8)$, and $G(9, 8)$. How is $\triangle DFG$ related to $\triangle ABC$?

_____

3. Draw the triangle formed by $A(0,2)$, $K(0, 6)$, and $D(6, 6)$. How is $\triangle AKD$ related to $\triangle ABC$?

_____

4. What is true about the hypotenuses of the three triangles in Steps 1, 2, and 3?

_____

### Essential Question

HOW can you determine congruence and similarity?

**Common Core State Standards**

**Content Standards**
8.EE.6

**MP Mathematical Practices**
1, 2, 3, 4

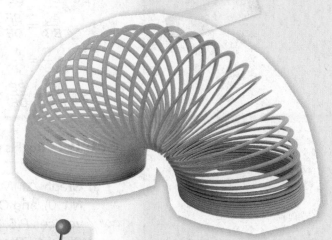

**Which MP Mathematical Practices did you use?**
**Shade the circle(s) that applies.**

① Persevere with Problems
② Reason Abstractly
③ Construct an Argument
④ Model with Mathematics
⑤ Use Math Tools
⑥ Attend to Precision
⑦ Make Use of Structure
⑧ Use Repeated Reasoning

# Similar Triangles and the Coordinate Plane

In the figure shown, △ABC and △BDE are slope triangles. Slope triangles are similar.

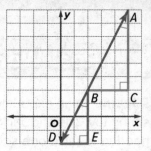

| | |
|---|---|
| ∠BAC ≅ ∠DBE | Given |
| ∠ACB ≅ ∠BED | Given |
| △ABC ~ △BDE | Angle-Angle Similarity |

You can use the properties of similar triangles to show the ratios of the rise to the run for each right triangle are equal.

## Example

Tutor

**1.** Write a proportion comparing the rise to the run for each of the similar slope triangles shown above. Then find the numeric value.

$$\frac{AC}{BE} = \frac{BC}{DE}$$   Corresponding sides of similar triangles are proportional.

$$AC \cdot DE = BE \cdot BC$$   Find the cross products.

$$\frac{AC \cdot DE}{BC \cdot DE} = \frac{BE \cdot BC}{BC \cdot DE}$$   Division Property of Equality

$$\frac{AC}{BC} = \frac{BE}{DE}$$   Simplify.

$$\frac{6}{3} = \frac{4}{2}$$   $AC = 6, BC = 3, BE = 4, DE = 2$

So, $\frac{AC}{BC} = \frac{BE}{DE}$, or $\frac{6}{3} = \frac{4}{2}$.

Show your work.

**Got it?** Do this problem to find out.

a. _____

a. Graph △MNO with vertices M(3, 1), N(1, 0), and O(3, 0), and △PQR with vertices P(5, 2), Q(−1, −1), and R(5, −1). Then write a proportion comparing the rise to the run for each of the similar slope triangles and find the numeric value.

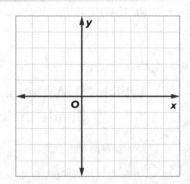

# Similar Triangles and Slope

**Words** The ratio of the rise to the run of two slope triangles formed by a line is equal to the slope of the line.

**Example**

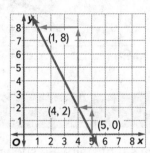

(1, 8)
(4, 2)
(5, 0)

**Larger Triangle**
$$\frac{\text{rise}}{\text{run}} = \frac{6}{-3}, \text{ or } -2$$

**Smaller Triangle**
$$\frac{\text{rise}}{\text{run}} = \frac{2}{-1}, \text{ or } -2$$

$$\text{slope} = \frac{-2}{1}, \text{ or } -2$$

The ratios of the rise to the run of the two similar slope triangles in Example 1 are the same as the slope of the line. Since the ratios are equal, the slope $m$ of a line is the same between any two distinct points on a non-vertical line in the coordinate plane.

# Example

Tutor

**2.** The pitch of a roof refers to the slope of the roof line. Choose two points on the roof and find the pitch of the roof shown. Then verify that the pitch is the same by choosing a different set of points.

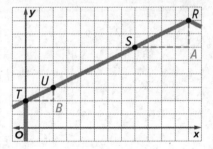

$$m = \frac{y_2 - y_1}{x_2 - x_1} \quad \text{Formula for slope}$$

$$m = \frac{8 - 6}{12 - 8} \quad \text{Use the points } S \text{ and } R. \ (x_1, y_1) = (8, 6) \text{ and } (x_2, y_2) = (12, 8)$$

$$m = \frac{2}{4} \text{ or } \frac{1}{2} \quad \text{Simplify.}$$

The pitch of the roof is $\frac{1}{2}$. Verify that the pitch is the same using two other points.

$$m = \frac{y_2 - y_1}{x_2 - x_1} \quad \text{Formula for slope}$$

$$m = \frac{2 - 3}{0 - 2} \quad \text{Use the points } U \text{ and } T. \ (x_1, y_1) = (2, 3) \text{ and } (x_2, y_2) = (0, 2)$$

$$m = \frac{-1}{-2} \text{ or } \frac{1}{2} \quad \text{Simplify. The pitch is the same.}$$

**STOP** and Reflect

Is the statement
$\triangle RAS \sim \triangle UBT$ true?
Explain below.

**Got it?** Do this problem to find out.

b. _____

b. The plans for a teeter-totter are shown at the right. Using points G and L, find the slope of the teeter-totter. Then verify that the slope is the same at a different location by choosing a different set of points.

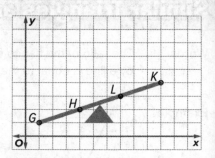

# Guided Practice

Check

1. Graph △ACG with vertices A(1, 4), C(3, −2), and G(1, −2), and △BCF with vertices B(2, 1), C(3, −2), and F(2, −2). Then write a proportion comparing the rise to the run for each of the similar slope triangles and find the numeric value. (Example 1)

_____

2. The plans for a set of stairs are shown below. Using points X and Z, find the slope of the line down the stairs. Then verify that the slope is the same at a different location by choosing a different set of points. (Example 2)

_____

Show your work.

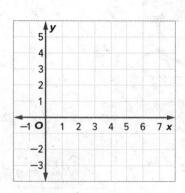

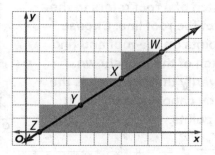

3.  **Building on the Essential Question** How is the slope of a line related to the similar slope triangles formed by the line?

_____

_____

_____

_____

_____

**Rate Yourself!**

How confident are you about slope and similar triangles? Check the box that applies.

☹  😐  🙂

☐ ☐ ☐ ☐ ☐

For more help, go online to access a Personal Tutor.

Tutor 💬

# Independent Practice

Go online for Step-by-Step Solutions

**Graph each pair of similar triangles. Then write a proportion comparing the rise to the run for each of the similar slope triangles and find the numeric value.** (Example 1)

**1** △ABC with vertices A(−6, −1), B(−4, −1), and C(−6, −3); △NLM with vertices N(−3, 3), L(0, 3), and M(−3, 0)

**2.** △FGH with vertices F(2, 3), G(2, −1), and H(−6, 3); △JKL with vertices J(0, 2), K(0, 0), and L(−4, 2)

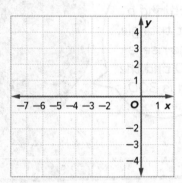

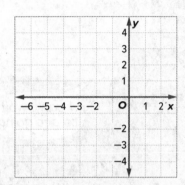

**3** The plans for a skateboard ramp are shown. Use two points to find the slope of the ramp. Then verify that the slope is the same at a different location by choosing a different set of points. (Example 2)

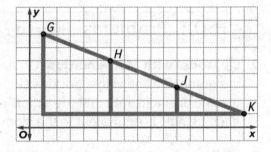

**4.** A ladder is leaning up against the side of a house. Use two points to find the slope of the ladder. Then verify that the slope is the same at a different location by choosing a different set of points.

(Example 2)

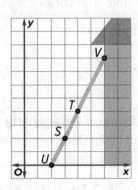

**5.** **MP** **Reason Abstractly** Triangle XYZ has vertices X(0, 0), Y(10, 0), and Z(0, 6). Triangle MYP has vertices M(5, 0), Y(10, 0), and P(x, y). Find the missing coordinates for P if △MYP ~ △XYZ.

6. **MP Model with Mathematics** Refer to the graphic novel frame below. On the beach, a cable is attached to the pier. The line formed by the cable has a slope of $\frac{3}{5}$. Is the triangle formed by the pier, the beach, and the cable similar to the triangle formed by the boat, the parasailer, and the rope? Explain. _____

_____

7. **MP Model with Mathematics** On a separate piece of grid paper, draw the graph of a line with a positive slope. Draw two slope triangles formed by the line. Demonstrate that the simplified ratio of the rise to the run of each triangle is equivalent to the slope.

8. **MP Persevere with Problems** The slope of a line is −3.5. Find two possible measurements for the legs of similar slope triangles. Explain your reasoning. _____

_____

_____

9. **MP Reason Inductively** Triangle *JKL* has vertices *J*(0, 0), *K*(1, 0), and *L*(1, 2). Determine if each triangle is similar to and/or a slope triangle with △*JKL*.

   a. △*ABC*: *A*(1, 2), *B*(1, 6), *C*(3, 6) _____

   b. △*MNP*: *M*(3, 1), *N*(6, 1), *P*(6, 7) _____

   c. △*RST*: *R*(1, 2), *S*(4, 2), *T*(4, 5) _____

   d. △*WXY*: *W*(0, 0), *X*(−1, −2), *Y*(0, −2) _____

# Extra Practice

**Graph each pair of similar triangles. Then write a proportion comparing the rise to the run for each of the similar slope triangles and find the numeric value.**

**10.** $\triangle LKM$ with vertices $L(-4, 4)$, $K(-4, -4)$, and $M(2, -4)$; $\triangle NPM$ with vertices $N(-1, 0)$, and $P(-1, -4)$

$$\frac{LK}{KM} = \frac{NP}{PM}, \text{ or } -\frac{4}{3}$$

Graph and label each triangle.

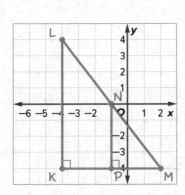

Write the proportion using the side labels.

$$\frac{LK}{KM} = \frac{NP}{PM}$$

$$-\frac{8}{6} = -\frac{4}{3}$$

**11.** $\triangle ABC$ with vertices $A(-5, -6)$, $B(1, -6)$, and $C(1, 3)$; $\triangle GFD$ with vertices $G(-3, -3)$, $F(-1, -3)$, and $D(-1, 0)$

_____

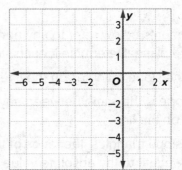

**MP** **Model with Mathematics** Use a graph to find the missing coordinates for point $Z$ if $\triangle MNP \sim \triangle XYZ$.

**12.** $M(-2, -3)$, $N(2, -3)$, $P(2, 3)$, $X(0, 0)$, $Y(2, 0)$

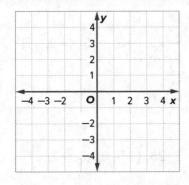

**13.** $M(5, 0)$, $N(5, -3)$, $P(2, -3)$, $X(7, 2)$, $Y(1, 2)$

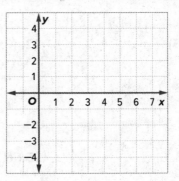

**Copy and Solve** Find the missing coordinates for point $D$ if $\triangle ABC \sim \triangle DEF$.
**Show your work on a separate sheet of paper.**

**14.** $A(-1, 3)$, $B(1, 3)$, $C(1, 6)$,
$E(-4, -7)$, $F(-4, -1)$

**15.** $A(1, 11)$, $B(1, 6)$, $C(3, 6)$,
$E(1, 1)$, $F(5, 1)$

**16.** Triangle *ABC* with vertices *A*(−3, 7), *B*(−3, 5), and *C*(0, 5), and triangle *CDE* with vertices *C*(0, 5), *D*(0, 1), and *E*(6, 1) are slope triangles.

Draw the triangles and the line that they represent on the coordinate plane.

Find the slope of the line. Then describe the relationship between the slope triangles and the slope of the line.

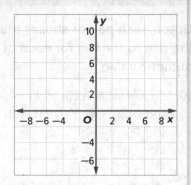

**17.** The statements below refer to any non-vertical line on the coordinate plane. Determine if each statement is true or false

   **a.** All of the slope triangles on the line are similar.    ☐ True ☐ False

   **b.** The slope is the same between any two distinct points on the line.    ☐ True ☐ False

   **c.** In the slope triangles, the ratios of the rise to the run are equal to the absolute value of the slope.    ☐ True ☐ False

## CCSS Common Core Spiral Review

**Find the slope of the line that passes through each pair of points.** 7.RP.2

**18.** (2, 2), (−2, −2) _____

**19.** (5, −4), (9, −4) _____

**20.** (4, 3), (−1, 6) _____

**21.** (3, 3), (3, 5) _____

**22.** (0, 0), (3, −6) _____

**23.** (−8, −15), (−2, −5) _____

**24.** (−3, 5), (3, 6) _____

**25.** (0.2, 0.7), (1.7, 1.2) _____

**26.** (−5, 0), (3, −2) _____

# Area and Perimeter of Similar Figures

 **Real-World Link**

**Games** Four square is a ball game played on a hard surface. The court is a 16-foot by 16-foot square divided into four equal squares.

 **Essential Question**

HOW can you determine congruence and similarity?

**CCSS** **Common Core State Standards**

**Content Standards**
Extension of 8.G.4

**MP Mathematical Practices**
1, 2, 3, 4

1. Use the figure to draw a four square court. Divide each side in half. Draw lines to divide the court into four equal squares. Is each smaller square similar or congruent to the larger square? Explain. _____

_____

_____

2. What is the perimeter of the larger square drawn above?

the smaller square? ☐ centimeters; ☐ centimeters

3. How is the perimeter of one of the smaller squares related to the perimeter of the larger square and the scale factor?

_____

_____

**Which MP Mathematical Practices did you use?**
**Shade the circle(s) that applies.**

① Persevere with Problems
② Reason Abstractly
③ Construct an Argument
④ Model with Mathematics

⑤ Use Math Tools
⑥ Attend to Precision
⑦ Make Use of Structure
⑧ Use Repeated Reasoning

# Perimeter and Area of Similar Figures

Work Zone

**Perimeter**

**Words**   If figure $B$ is similar to figure $A$ by a scale factor, then the perimeter of $B$ is equal to the perimeter of $A$ times the scale factor.

**Symbols**
$$\text{perimeter of figure } B = \text{perimeter of figure } A \cdot \text{scale factor}$$

**Models**

**Figure A**

**Area**

**Words**   If figure $B$ is similar to figure $A$ by a scale factor, then the area of $B$ is equal to the area of $A$ times the square of the scale factor.

**Symbols**
$$\text{area of figure } B = \text{area of figure } A \cdot (\text{scale factor})^2$$

**Figure B**

In similar figures, the perimeters are related by the scale factor, $k$. What about area? The area of one similar figure is equal to the area of the other similar figure times the *square* of the scale factor, or $k^2$.

Tutor

## Example

1. **Two rectangles are similar. One has a length of 6 inches and a perimeter of 24 inches. The other has a length of 7 inches. What is the perimeter of this rectangle?**

   The scale factor is $\frac{7}{6}$. The perimeter of the original is 24 inches.

   $x = 24\left(\frac{7}{6}\right)$      Multiply by the scale factor.

   $x = \overset{4}{\frac{24}{1}}\left(\frac{7}{\underset{1}{6}}\right)$      Divide out common factors.

   $x = 28$      Simplify.

   So, the perimeter of the new rectangle is 28 inches.

*Show your work.*

**Got it?**  **Do this problem to find out.**

a. _____

a. Triangle $LMN$ is similar to triangle $PQR$. If the perimeter of $\triangle LMN$ is 64 meters, what is the perimeter of $\triangle PQR$?

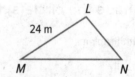
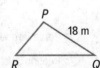

# Example

Tutor

**2.** In a scale drawing, the perimeter of the garden is **64** inches. The actual length of $\overline{AB}$ is 18 feet. What is the perimeter of the actual garden?

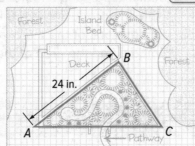

**Step 1** The actual length is proportional to the length in the drawing with a ratio of $\frac{18 \text{ ft}}{24 \text{ in.}}$. Find the scale factor.

$$\frac{18 \text{ ft}}{24 \text{ in.}} = \frac{216 \text{ in.}}{24 \text{ in.}} \text{ or } \frac{9}{1}$$     Convert feet to inches and divide out units.

**Step 2** Find the perimeter of the actual garden.

perimeter of garden = perimeter of drawing • scale factor

$$P = 64 \cdot 9 \text{ or } 576$$     Substitute. Then simplify.

The perimeter of the actual garden is 576 inches or 48 feet.

---

**Got it?** Do this problem to find out.

Show your work.

**b.** Two quilting squares are shown. The scale factor is 3:2. What is the perimeter of square *TUVW*?

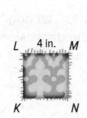 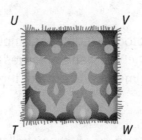

b. _____

---

Tutor

# Example

**3.** The Eddingtons have a 5-foot by 8-foot porch on the front of their house. They are building a similar porch on the back with double the dimensions. Find the area of the back porch.

The scale factor is 2.

The area of the front porch is (5)(8) or 40 square feet.

$$x = 40(2)^2$$     Multiply by the square of the scale factor.

$$x = 40(4) \text{ or } 160$$     Evaluate the power.

The back porch will have an area of 160 square feet.

**Got it?** Do this problem to find out.

c. _____

**c.** Malia is painting a mural on her bedroom wall. The image she is reproducing is 4.8 inches by 7.2 inches. If the dimensions of the mural are 10 times the dimensions of the image, find the area of the mural in square inches.

## Guided Practice

Check ✓

**For each pair of similar figures, find the perimeter of the second figure.**
(Example 1)

**1.**

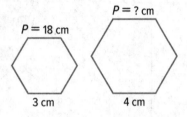

$P = 18$ cm    $P = ?$ cm

3 cm    4 cm

**2.**

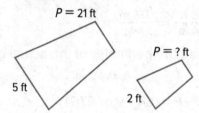

$P = 21$ ft    $P = ?$ ft

5 ft    2 ft

**3.** Julie is enlarging a digital photograph on her computer. The original photograph is 5 inches by 7 inches. If she enlarges the dimensions 1.5 times, what will be the perimeter and area of the new image? (Examples 2 and 3)

**4.** Logan is flying a kite that is made up of three similar rectangles. The sides of the three rectangles are in the ratio 1:2:3. If the area of the smallest rectangle is 72 square inches, what are the areas of the other two rectangles? (Example 3) _____

**5.** ⓔ **Building on the Essential Question** If you know two figures are similar and you are given the area of both figures, how can you determine the scale factor of the similarity?

**Rate Yourself!**

☐ I understand how to find the perimeter and area of similar figures.

▶▶ Great! You're ready to move on!

☐ I still have some questions about the perimeter and area of similar figures.

📖 No Problem! Go online to access a Personal Tutor.  Tutor

Name _____  My Homework _____

**For each pair of similar figures, find the perimeter of the second figure.**
(Example 1)

**1**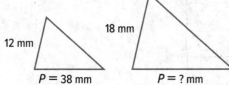

12 mm    18 mm

P = 38 mm          P = ? mm

**2.**

8.4 in.          6.3 in.

P = 19.4 in.          P = ? in.

**3.** The city of Brice is planning to build a skate park. An architect designed the area shown at the right. In the plan, the perimeter of the park is 80 inches. If the actual length of $\overline{WX}$ is 50 feet, what will be the perimeter of the actual skate

park? (Example 2) _____

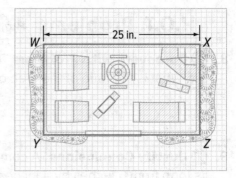

W |————— 25 in. —————| X

Y                          Z

**4.** A child's desk is made so that the dimensions are two-thirds the dimensions of a full-size adult desk. Suppose the top of the full-size desk measures 54 inches long by 36 inches wide. What is the perimeter and area of the top of the child's desk? (Examples 2 and 3)

_____

**5** Theo is constructing a miniature putting green in his backyard. He wants it to be similar to a putting green at the local golf course, but one third the dimensions. The area of the putting green at the golf course is 1,134 square feet. What will be the area of the putting green Theo constructs?

_____

**6.** Craig is making a model version of his neighborhood that uses model trains. The ratio of the model train to the actual train is 1:64. His neighborhood covers an area of 200,704 square feet. What will be the area of the model neighborhood?

_____

7. **MP Identify Structure** Complete the graphic organizer to compare how the scale factor affects the side lengths, perimeter, and area of similar rectangles.

| If the scale factor is... | Multiply the ... | | | |
|---|---|---|---|---|
| | Length by | Width by | Perimeter by | Area by |
| 2 | | | | |
| 4 | | | | |
| 0.5 | | | | |
| $\frac{2}{3}$ | | | | |
| k | | | | |

## H.O.T. Problems  Higher Order Thinking

8. **MP Persevere with Problems** Two circles have circumferences of $\pi$ and $3\pi$. What is the ratio of the area of the circles? the diameters? the radii?

_____

9. **MP Justify Conclusions** A company wants to reduce the dimensions of its logo from 6 inches by 4 inches to 3 inches by 2 inches to use on business cards. Robert thinks that the new logo is $\frac{1}{4}$ the size of the original logo. Denise thinks that is $\frac{1}{2}$ of the original size. Explain their thinking to a classmate. _____

_____

_____

10. **MP Use Math Tools** Use the coordinate plane to draw a rectangle. Dilate the rectangle and draw the dilation. Then determine the perimeter and area of each rectangle to model the effect of the dilation.

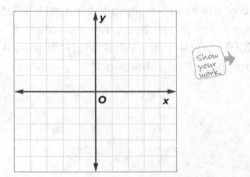

Show your work.

# Extra Practice

**For each pair of similar figures, find the unknown perimeter.**

**11.**

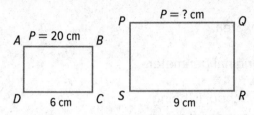

The scale factor is $\frac{3}{2}$. Multiply the

perimeter of ABCD by $\frac{3}{2}$.

$P = 20 \cdot \frac{3}{2}$ or 30

30 cm

**12.**

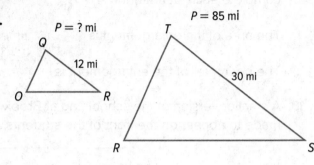

**13.** For your birthday party, you make a map to your house on a 3-inch-wide by 5-inch-long index card. What will be the perimeter and area of your map if you use a copier to enlarge it so it is 8 inches long?

_____

**14.** A company wants to reduce the dimensions of its logo by one fourth to use on business cards. If the area of the original logo is 4 square inches, what is the area of the logo that will be used on the business cards?

_____

**15.** Two picture frames are similar. The ratio of the perimeters of the two pieces is 3:5. If the area of the smaller frame is 108 square inches, what

is the area of the larger frame? _____

**16.** **MP** **Persevere with Problems** Mr. James is enlarging a logo for printing on the back of a T-shirt. He wants to enlarge a logo that is 3 inches by 5 inches so that the dimensions are 3 times larger than the original. How many times as large as the original logo will the area of the printing be?

_____

**17.** A photograph is enlarged to 3 times the size of the original. Fill in the boxes to complete each statement.

The area of the enlargement is ⬚ times the original area.

The perimeter of the enlargement is ⬚ times the original perimeter.

**18.** A smaller version of the school flag at Brook Park Middle School is being made to appear on the front of the students' homework agenda books.

5 ft

BROOK PARK EAGLES

3 ft

The perimeter of the smaller version of the flag is 2 feet. Select the correct values to complete each statement.

| 1 | 2 | 4 | 8 | 16 |
|---|---|---|---|---|
| $\frac{1}{64}$ | $\frac{1}{8}$ | $\frac{15}{8}$ | $\frac{15}{64}$ | |

**a.** The perimeter of the full size flag is ⬚ feet.

**b.** The scale factor of the reduction is ⬚ .

**c.** The area of the smaller version of the flag is ⬚ square feet.

CCSS **Common Core Spiral Review**

**Graph each figure with the given vertices and its image after the indicated transformation.** 8.G.3

**19.** △ABC: A(0, −1), B(0, 3), C(3, 3)
90° clockwise rotation about the origin

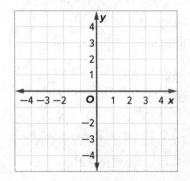

**20.** $\overline{XY}$: X(1, 1), Y(−2, −3)
translation of 1 unit right and 3 units up

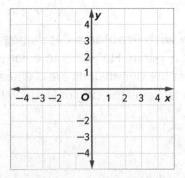

# 21ST CENTURY CAREER
## in Car Design

## Car Designer

Do you like drawing? Are you technical and precise in your drawings? You should consider a career as a car designer. Car designers use Computer Aided Design to create technical drawings that are used in manufacturing and construction. Information from architects and engineers is used to create highly specialized drawings that show how to construct everything from a nightstand to the space shuttle.

**College & Career READINESS**

Explore college and careers at ccr.mcgraw-hill.com

## Is This the Career for You?

Are you interested in a career as a car designer? Take some of the following courses in high school.

◆ Geometry
◆ Mechanical Drawing
◆ Computer Graphics
◆ Design

Turn the page to find out how math relates to a career in Car Design.

## MP Drive Yourself to Success

Use the information on the drawing to solve each problem.

1. What transformation maps the drawing to the actual car? _____

2. Are the views of the drawing of the car similar to views of the actual car? Explain.

_____

_____

3. If the scale factor is $\frac{1}{25}$, find the following:

a. the length of the actual car _____

b. the distance from the front wheel to the rear wheel of the actual car _____

4. If the actual height of the car is 60 inches, what is y? _____

5. If $x = 2\frac{4}{5}$ inches, what is the actual distance between the tires on the car?

_____

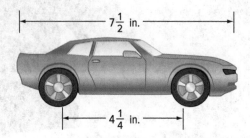

$7\frac{1}{2}$ in.

$4\frac{1}{4}$ in.

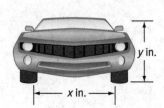

y in.

x in.

## MP Career Project

It's time to update your career portfolio! Describe the features that you, as a car designer, would include in a new car design. Determine whether these features already exist in cars today.

_____

_____

_____

_____

_____

_____

List several challenges associated with this career.

• _____

• _____

• _____

• _____

• _____

# Chapter Review

## Vocabulary Check

Reconstruct the vocabulary word and definition from the letters under the grid. The letters for each column are scrambled directly under that column.

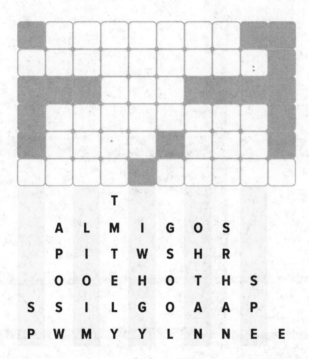

Complete each sentence using vocabulary from the chapter.

**1.** Two figures are _____ if one can be obtained from the other by a series of rotations, reflections, or translations.

**2.** _____ uses properties of similar polygons to find distances or lengths that are difficult to measure directly.

**3.** The parts of congruent figures that match are called

_____.

**4.** Two figures are _____ if one can be obtained from the other by a series of transformations and dilations.

**5.** When a transformation is applied to a figure and then another transformation is applied to the image, the result is called

a _____.

## Use Your FOLDABLES

**Use your Foldable to help review the chapter.**

Tape here

**Tab 1**      **Congruent Figures**

Draw

Draw

Draw

Draw

**Tab 2**      **Similar Figures**

Tape here

## Got it?

**Triangle *ABC* has vertices *A*(0, 0), *B*(2, 4), *C*(6, 0). Match each image with the description of its transformation.**

**1.** $A'(0, 0)$, $B'(2, -4)$, $C'(6, 0)$

**2.** $A'(0, 0)$, $B'(1, 2)$, $C'(3, 0)$

**3.** $A'(0, 0)$, $B'(4, -2)$, $C'(0, -6)$

**4.** $A'(2, -6)$, $B'(6, 2)$, $C'(14, -6)$

**a.** similar; a dilation with a scale factor of $\frac{1}{2}$

**b.** congruent; a 90° clockwise rotation about the origin

**c.** congruent; a reflection over the *x*-axis

**d.** similar; a translation of $(x + 1, y - 3)$ followed by a dilation with a scale factor of 2

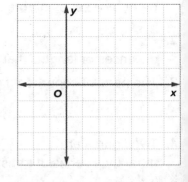

# Power Up! Performance Task

## Can Triangles Model Maps?

Andy is investigating the relationship between triangles and how they can model a city street map. On the graph, one unit is equal to one inch. The grid shows two right triangles. $\overline{DG} \cong \overline{CB}$, $m\angle B = 20.6°$, and $m\angle D = 69.4°$.

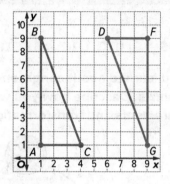

**Write your answers on another piece of paper. Show all of your work to receive full credit.**

### Part A

Are the two triangles congruent? Explain your reasoning. What transformation(s) could be used to help determine that the two triangles are congruent?

### Part B

The triangle formed by the intersections of Oak Road, Main Street, and Highway 33 is similar to triangle $ABC$. The section of Main Street between Oak Road and Highway 33 is eight miles long. Based on the information given, what is the length of Oak Road from Main Street north to the intersection of Highway 33? Set up a proportion and solve. Round your answer to the nearest tenth.

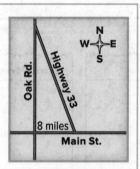

# Reflect

 **Answering the Essential Question**

Use what you learned about congruence and similarity to complete the graphic organizer. Describe how you would show congruence or similarity using measurements and transformations.

**Essential Question**

**HOW can you determine congruence and similarity?**

| Congruence | Similarity |
|---|---|
| **Definition** | **Definition** |

| Measurements | Transformations | Measurements | Transformations |
|---|---|---|---|
| | | | |

**Answer the Essential Question.** HOW can you determine congruence and similarity?

_____

_____

_____

# Chapter 8

# Volume and Surface Area

## Essential Question

WHY are formulas important in math and science?

## Common Core State Standards

**Content Standards**
8.G.9

 **Mathematical Practices**
1, 2, 3, 4, 6, 7

## Math in the Real World

**Ice Skating** During the winter, D'shaun and her friends watch speed skating races at a local park. The ice skating rink is made up of two semi-circles and a rectangle. What is the area of the rink?

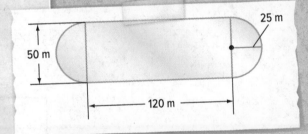

**Study Organizer**

Cut out the Foldable on page FL9 of this book.

Place your Foldable on page 652.

Use the Foldable throughout this chapter to help you learn about volume and surface area.

## Vocabulary

| | | |
|---|---|---|
| composite solids | lateral area | sphere |
| cone | nets | total surface area |
| cylinder | polyhedron | volume |
| hemisphere | similar solids | |

## Review Vocabulary

**Area** The area of a geometric figure is the measure of the surface enclosed by the figure. Write the correct area formula in each shape.

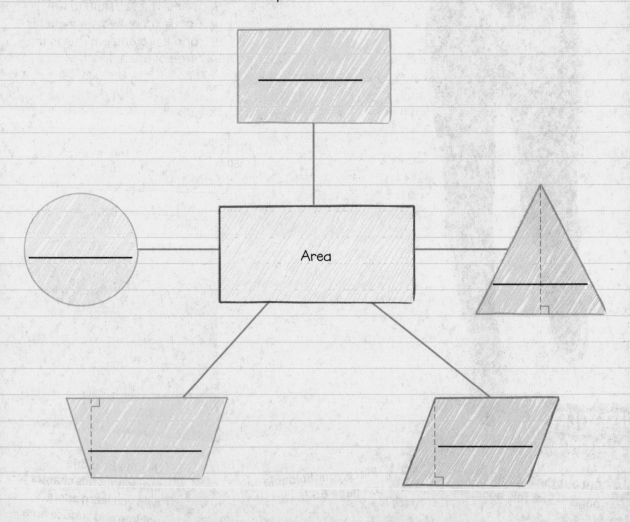

Read each statement. Decide whether you agree (A) or disagree (D). Place a checkmark in the appropriate column and then justify your reasoning.

| Volume and Surface Area | | | |
|---|---|---|---|
| Statement | A | D | Why? |
| To find the volume of a cylinder, multiply the area of the base by the height of the cylinder. | | | |
| The formula to find the volume of a cone is the same as the formula to find the volume of a cylinder. | | | |
| A sphere is a set of all points in space that are a given distance from a given point called the center. | | | |
| The lateral area of a three-dimensional figure is the sum of the areas of all its surfaces. | | | |
| The net of a cylinder consists of two circles and a rectangle. | | | |
| To find the lateral area of a cone, you multiply π by the radius of the base by the height of the cone. | | | |

## When Will You Use This?

Here are a few examples of how unit rates are used in the real world.

**Activity 1** Adrienne's mom made soup in a large pot and is going to pour it into quart jars. Name some ways her mom could estimate how many jars she would need.

_____

_____

**Activity 2** Go online at **connectED.mcgraw-hill.com** to read the graphic novel **Material Mayhem**. What do Danielle and Dion need to buy? _____

Danielle and Dion in
*Material Mayhem*

This works, but how much do we need?

Try the Quick Check below.
Or, take the Online Readiness Quiz.

Check ✓

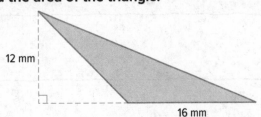
**Common Core Review** 7.G.6

## Example 1

Find the area of the triangle.

12 mm

16 mm

$A = \frac{1}{2}bh$      Formula for area of a triangle

$A = \frac{1}{2} \cdot 16 \cdot 12$      Replace b with 16 and h with 12.

$A = 96$      Simplify.

The area is 96 square millimeters.

## Example 2

Evaluate $\pi \cdot 16^2$. Use 3.14 for $\pi$. Round to the nearest tenth.

$\pi \cdot 16^2 \approx 3.14 \cdot 256$      Evaluate $16^2$.

$\approx 803.8$      Multiply.

### Quick Check

**Area** Find the area of each figure.

1.
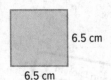
8 cm
17 cm

Show your work.

2.
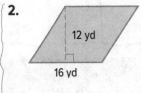
12 yd
16 yd

3.
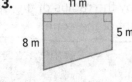
11 m
8 m
5 m

4.
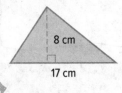
6.5 cm
6.5 cm

5.
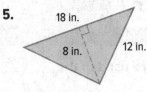
18 in.
8 in.
12 in.

6.
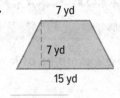
7 yd
7 yd
15 yd

**Evaluate** Find the value of each expression. Use 3.14 for $\pi$. Round to the nearest tenth.

7. $\pi \cdot 15 \approx$ _____

8. $2 \cdot \pi \cdot 3.2 \approx$ _____

9. $\pi \cdot (19 \div 2)^2 \approx$ _____

### How Did You Do?

**Which problems did you answer correctly in the Quick Check?**
**Shade those exercise numbers below.**

①  ②  ③  ④  ⑤  ⑥  ⑦  ⑧  ⑨

# Inquiry Lab
## Three-Dimensional Figures

 **Inquiry** **HOW are some three-dimensional figures related to circles?**

 **CCSS** **Content Standards** Preparation for 8.G.9

**MP** **Mathematical Practices** 1, 3

Kenji is training his dog to run through an agility course. One of the activities in the course is the tunnel, an open tube through which the dog runs.

## Hands-On Activity

A three-dimensional figure with faces that are polygons is called a **polyhedron**. There are three-dimensional figures that are *not* polyhedrons. Some examples of these figures are *cylinders*, *cones*, and *spheres*.

**Step 1** For each figure, list three real-world items that represent the figure.

| Cylinder | Cone | Sphere |
|---|---|---|
|  |  |  |
|  |  |  |
|  |  |  |

**Step 2** Just as a rectangular prism and a pyramid have bases, a cylinder and a cone have bases as well. What is the shape of the base of a cylinder?

_____ a cone? _____

**Step 3** Interesting shapes can occur when you find the cross section of a figure that is not a polyhedron. Describe the shape of the figure resulting from a horizontal cross section of each of the following.

_____  _____

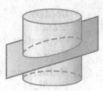 

Collaborate

**Work with a partner. Draw and describe the shape resulting from each cross section.**

1.

2.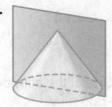

3.

 Show your work.

4.

5.

6.

## Create

On Your Own

7. **MP Use a Counterexample** *True* or *false*: The cross section of a cylinder, a cone, and a sphere will *always* be a circle or an oval. If false, provide a counterexample.

_____

_____

8. **Inquiry** HOW are some three-dimensional figures related to circles?

_____

_____

# Volume of Cylinders

 **Real-World Link**

**Jelly Beans** Olivia's teacher filled a cylindrical jar with jelly beans. She is awarding a prize to the student who most accurately estimates the number of jelly beans in the jar. Olivia used a soup can to model the jar and centimeter cubes to model the jelly beans.

 **Work with a partner.**
Collaborate

1. Set a soup can on a piece of grid paper. Trace the area around the base as shown.

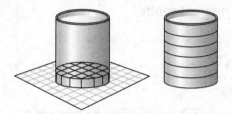

About how many centimeter cubes would fit at the bottom of the container? Remember to include partial cubes in your total. _____

2. Suppose each layer is 1 centimeter high. How many layers would it take to fill the cylinder? ☐

3.  **Be Precise** Write a formula that allows you to find the volume of the container. _____

**Which MP Mathematical Practices did you use?**
**Shade the circle(s) that applies.**

① Persevere with Problems
② Reason Abstractly
③ Construct an Argument
④ Model with Mathematics
⑤ Use Math Tools
⑥ Attend to Precision
⑦ Make Use of Structure
⑧ Use Repeated Reasoning

 **Essential Question**

WHY are formulas important in math and science?

 **Vocabulary**

volume
cylinder
composite solids

 **Common Core State Standards**

Content Standards
8.G.9
MP Mathematical Practices
1, 3, 4, 6

# Volume of a Cylinder

| Words | The volume $V$ of a cylinder with radius $r$ is the area of the base $B$ times the height $h$. | **Model** |
|---|---|---|

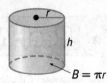

| Symbols | $V = Bh$, where $B = \pi r^2$ or $V = \pi r^2 h$ |
|---|---|

$B = \pi r^2$

**Work Zone**

**STOP and Reflect**

What formula do you use to find the area of the base of a cylinder?

**Volume** is the measure of the space occupied by a solid. Volume is measured in cubic units. A **cylinder** is a three-dimensional figure with two parallel congruent circular bases connected by a curved surface. The area of the base of a cylinder tells the number of cubic units in one layer. The height tells how many layers there are in the cylinder.

## Examples

Watch Tutor

**1.** **Find the volume of the cylinder. Round to the nearest tenth.**

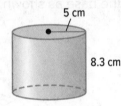

5 cm

8.3 cm

$V = \pi r^2 h$   Volume of a cylinder

$V = \pi(5)^2(8.3)$   Replace $r$ with 5 and $h$ with 8.3.

Use a calculator.

2nd [π] × 5 $x^2$ × 8.3 ENTER 651.8804756

The volume is about 651.9 cubic centimeters.

**Circles**

Recall that the radius is half the diameter.

**2.** **Find the volume of a cylinder with a diameter of 16 inches and a height of 20 inches. Round to the nearest tenth.**

$V = \pi r^2 h$   Volume of a cylinder

$V = \pi(8)^2(20)$   The diameter is 16 so the radius is 8. Replace $h$ with 20.

$V \approx 4,021.2$   Use a calculator.

The volume is about 4,021.2 cubic inches.

**Got it?** Do these problems to find out.

Show your work.

**Find the volume of each cylinder. Round to the nearest tenth.**

a. _____

a.

3 in.

1.8 in.

b.   diameter: 12 mm

height: 5 mm

b. _____

## Example

Tutor

**3.** A metal paperweight is in the shape of a cylinder. The paperweight has a height of 1.5 inches and a diameter of 2 inches. How much does the paperweight weigh if 1 cubic inch weighs 1.8 ounces? Round to the nearest tenth.

First find the volume of the paperweight.

| $V = \pi r^2 h$ | Volume of a cylinder |
| $V = \pi (1)^2 1.5$ | Replace $r$ with 1 and $h$ with 1.5. |
| $V \approx 4.7$ | Simplify. |

To find the weight of the paperweight, multiply the volume by 1.8.

$4.7(1.8) = 8.46$

So, the weight of the paperweight is about 8.5 ounces.

**Got it?** Do this problem to find out.

Show your work.

**c.** The Roberts family uses a container shaped like a cylinder to recycle aluminum cans. It has a height of 4 feet and a diameter of 1.5 feet. The container is full. How much do the contents weigh if the average weight of aluminum cans is 37 ounces per cubic foot? Round to the nearest tenth.

c. _____

# Volume of a Composite Solid

Objects made up of more than one type of solid are called **composite solids.** To find the volume of a composite solid, decompose the figure into solids whose volumes you know how to find.

## Example

Tutor

**4.** Tanya uses cube-shaped beads to make jewelry. Each bead has a circular hole through the middle. Find the volume of each bead.

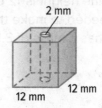
2 mm
12 mm
12 mm
12 mm

The bead is made of one rectangular prism and one cylinder. Find the volume of each solid. Then subtract to find the volume of the bead.

**Rectangular Prism**

$V = Bh$

$V = (12 \cdot 12)12$ or 1,728

**Cylinder**

$V = Bh$

$V = (\pi \cdot 1^2)12$ or 37.7

The volume of the bead is $1,728 - 37.7$ or 1,690.3 cubic millimeters.

**Got it?** Do this problem to find out.

d. _____

**d.** The Service Club is building models of storage chests, like the one shown, to donate to a charity. Find the volume of the chest to the nearest tenth.

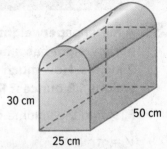

30 cm

50 cm

25 cm

Check ✓

# Guided Practice

**Find the volume of each cylinder. Round to the nearest tenth.** (Examples 1 and 2)

**1.** _____

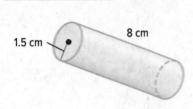

1.5 cm

8 cm

**2.** _____

diameter: 8 in.
height: 8 in.

**3.** A platform like the one shown was built to hold a sculpture for an art exhibit. What is the volume of the figure? (Example 4)

_____

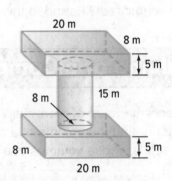

20 m

8 m

5 m

8 m

15 m

8 m

5 m

20 m

**4.** A scented candle is in the shape of a cylinder. The radius is 4 centimeters and the height is 12 centimeters. Find the mass of the wax needed to make the candle if 1 cubic centimeter of wax has a mass of 3.5 grams. Round to the nearest tenth. (Example 3)

_____

**5.**  **Building on the Essential Question** How is the formula for the volume of a cylinder similar to the formula for the volume of a rectangular prism?

_____

_____

**Rate Yourself!**

How confident are you about volume of cylinders? Check the box that applies.

For more help, go online to access a Personal Tutor.

Tutor

**FOLDABLES** Time to update your Foldable!

Name _____  My Homework _____

# Extra Practice

Copy and Solve For Exercises 10–27, show your work and answers on a
separate piece of paper.

**Find the volume of each cylinder. Round to the nearest tenth.**

**10.**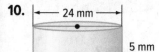

**11.**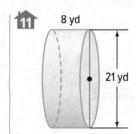

**12.**

**13.** Kyle has a container of flour in the shape of a cylinder. The container has
a diameter of 10 inches and a height of 8 inches. If the container is full,
how much will the flour weigh if the average weight of flour is 0.13 ounces
per cubic inch? Round to the nearest tenth.

**14.** Charlotte wants to make a mailbox like the one shown. What is the
volume of the mailbox? Round to the nearest tenth.

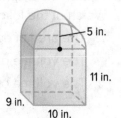

**15.** Cylinder A has a radius of 4 inches and a height of 2 inches. Cylinder B
has a radius of 2 inches. What is the height of Cylinder B to the nearest
inch if both cylinders have the same volume?

**16.** Which will hold more cake batter, the rectangular pan or two round pans?
Explain your reasoning to a classmate.

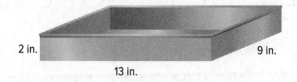 

**17.** (MP) **Multiple Representations** The dimensions for
four cylinders are shown in the table.

a. **Symbols** Write an equation to find the volume
of each cylinder.

b. **Words** Compare the dimensions of Cylinder A
with the dimensions of Cylinders B, C, and D.

c. **Numbers** Complete the table.

d. **Words** Explain how changing the dimensions of
a cylinder affects the cylinder's volume.

| | Radius (cm) | Height (cm) | Volume (cm³) |
|---|---|---|---|
| **Cylinder A** | 1 | 1 | |
| **Cylinder B** | 1 | 2 | |
| **Cylinder C** | 2 | 1 | |
| **Cylinder D** | 2 | 2 | |

**18.** Without doing any calculations, do you think Cylinder 1 and Cylinder 2 will have the same volume? Explain your reasoning.

**Cylinder 1**

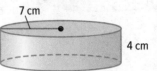

Fill in each box to complete the following statements.

To the nearest tenth, the volume of Cylinder 1 is _____ .

To the nearest tenth, the volume of Cylinder 2 is _____ .

**Cylinder 2**

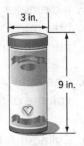

**19.** The oatmeal container shown has a diameter of 3 inches and a height of 9 inches. Which of the following statements are true? Select all that apply.

☐ The area of each base is exactly $9\pi$ square inches.

☐ The volume of the container is exactly $20.25\pi$ cubic inches.

☐ The volume of the container to the nearest tenth is about 63.6 cubic inches.

## Common Core Spiral Review

**Find the area of each circle. Round to the nearest tenth.** 7.G.4

**20.**

15 in.

**21.**

8 cm

**22.**
9 in.

**23.**
3 in.

**24.**

6.2 cm

**25.**
4 m

**Find the volume of each prism.** 7.G.6

**26.**

6 ft
2 ft   3 ft

**27.**
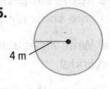
7 m   14 m
11 m

## Lesson 2

# Volume of Cones

 **Real-World Link**

**Carnival** Grace and Elle are making snow cones for the school carnival. They want to know how much ice goes into a paper cone that has a radius of 1.5 inches and a height of 4 inches.

1. Recall the formula for finding the volume of a rectangular pyramid is $V = \frac{1}{3}Bh$. How does the volume of a pyramid compare to the volume of a prism with the same base and height?

_____

2. What is the formula for finding the volume of a cylinder?

_____

3. What is the volume of a cylinder with a radius of 1.5 inches and a height of 4 inches? Use $\pi = 3.14$.

_____

4. The volume of the cones Grace and Elle are using is about 9.42 cubic inches. Write a ratio in simplest form comparing

the volume of the cone to the volume of the cylinder. $\dfrac{\boxed{\phantom{x}}}{\boxed{\phantom{x}}}$

5. 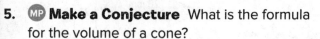 **Make a Conjecture** What is the formula for the volume of a cone?

_____

 **Essential Question**

WHY are formulas important in math and science?

**Vocab** **Vocabulary**

cone

 **Common Core State Standards**

**Content Standards**
8.G.9
**MP Mathematical Practices**
1, 2, 3, 4

**Which MP Mathematical Practices did you use?**
**Shade the circle(s) that applies.**

① Persevere with Problems  ⑤ Use Math Tools

② Reason Abstractly  ⑥ Attend to Precision

③ Construct an Argument  ⑦ Make Use of Structure

④ Model with Mathematics  ⑧ Use Repeated Reasoning

| Key Concept | Volume of a Cone |
| --- | --- |

| | | Model |
| --- | --- | --- |
| **Words** | The volume *V* of a cone with radius *r* is one third the area of the base *B* times the height *h*. |  |
| **Symbol** | $V = \frac{1}{3}Bh$ or $V = \frac{1}{3}\pi r^2 h$ | |

A **cone** is a three-dimensional figure with one circular base connected by a curved surface to a single vertex.

## Example

**1.** **Find the volume of the cone. Round to the nearest tenth.**

$V = \frac{1}{3}\pi r^2 h$ ............ Volume of a cone

$V = \frac{1}{3} \cdot \pi \cdot 3^2 \cdot 6$ ...... $r = 3, h = 6$

$V \approx 56.5$ ............ Simplify.

The volume is about 56.5 cubic inches.

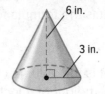

*Show your work.*

### Got it? Do these problems to find out.

Find the volume of each cone. Round to the nearest tenth.

a.

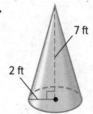

7 ft
2 ft

b.

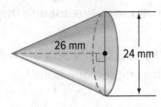

26 mm
24 mm

**Work Zone**

a. _____

b. _____

*Real World*

## Example

**2.** **A cone-shaped paper cup is filled with water. The height of the cup is 10 centimeters and the diameter is 8 centimeters. What is the volume of the paper cup? Round to the nearest tenth.**

$V = \frac{1}{3}\pi r^2 h$ ............ Volume of a cone

$V = \frac{1}{3} \cdot \pi \cdot 4^2 \cdot 10$ ...... $r = 4, h = 10$

$V \approx 167.6$ ............ Simplify.

The volume of the paper cup is about 167.6 cubic centimeters.

## Got it? Do this problem to find out.

c. April is filling six identical cones for her piñata. Each cone has a radius of 1.5 inches and a height of 9 inches. What is the total volume of the cones? Round to the nearest tenth.

c. _____

# Volume of Composite Solids

When a composite solid includes cylinders and cones, you can find the volume by decomposing it into solids whose volumes you know how to find.

## Example

**3.** Find the volume of the solid. Round to the nearest tenth.

**Step 1** Find the volume of the cylinder.

$V = \pi r^2 h$ — Volume of a cylinder

$V = \pi \cdot 4^2 \cdot 4$ — $r = 4, h = 4$

$V = \pi \cdot 16 \cdot 4$ — Simplify.

$V \approx 201.1$ — Simplify.

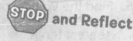
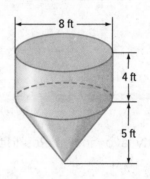

8 ft · 4 ft · 5 ft

**Step 2** Find the volume of the cone.

$V = \frac{1}{3}\pi r^2 h$ — Volume of a cone

$V = \frac{1}{3} \cdot \pi \cdot 4^2 \cdot 5$ — $r = 4, h = 5$

$V = \frac{1}{3} \cdot \pi \cdot 16 \cdot 5$ — Simplify.

$V \approx 83.8$ — Simplify.

So, the volume of the solid is about 201.1 + 83.8 or 284.9 cubic feet.

**STOP and Reflect**

Dario and Divya are simplifying $\pi \cdot 5^2$. Dario rounds $\pi$ to 3.14 and Divya uses the $\pi$ key on her calculator. Which student's calculation is closer to the exact value? Explain below.

## Got it? Do this problem to find out.

d. Find the volume of the solid.

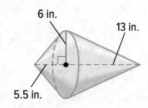

6 in. · 13 in. · 5.5 in.

d. _____

**Find the volume of each cone. Round to the nearest tenth.** (Examples 1 and 2)

**1.** _____

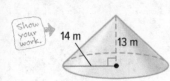

**2.** _____

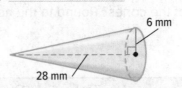

**3.** height: 9 m

diameter: 10 m _____

**4.** height: 120 millimeters

radius: 45 millimeters _____

**5.** Find the volume of the solid at the right. Round to the

nearest tenth. (Example 3) _____

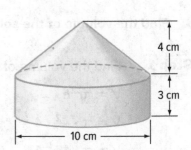

**6.** Find the volume of the party favor shown. Round to the

nearest tenth. (Example 3) _____

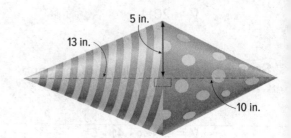

**7.** 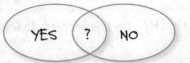 **Building on the Essential Question** What would
have a greater effect on the volume of a cone: doubling
its radius or doubling its height? Explain.

_____

_____

_____

_____

_____

## Rate Yourself!

How confident are you about
volume of cones? Shade the
section that applies.

YES   ?   NO

For more help, go online to
access a Personal Tutor.

Tutor

**FOLDABLES** Time to update your Foldable!

# Extra Practice

**Copy and Solve** For Exercises 15–33, show your work and answers on a separate piece of paper.

**Find the volume of each cone. Round to the nearest tenth.**

**15** 13.4 mm
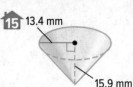
15.9 mm

**16.**
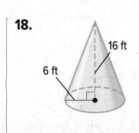
17 ft
4 ft

**17.**
9 m
6.5 m

**18.**
16 ft
6 ft

**19.** height: 24 centimeters
diameter: 8 centimeters

**20.** height: 9 inches
diameter: $7\frac{1}{2}$ inches

**21.** Austin is using the funnel shown to fill a glass bottle with colored sand. Estimate the volume of the funnel.

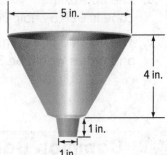
5 in.
4 in.
1 in.
1 in.

**22.** Mount Rainier, a cone-shaped volcano in Washington, is about 4.4 kilometers tall and about 18 kilometers across its base. Find the volume of Mount Rainier to the nearest whole number.

**23.** The volume of a cone is 471.24 cubic inches and the height is 8 inches. What is the diameter?

**24.** The volume of a cone is 593.46 cubic inches. The radius is 9 inches. Find the height of the cone to the nearest inch.

**MP Persevere with Problems** Find the height of each cone. Round to the nearest tenth.

**25.**

h   3 m
Volume: 42.39 m³

**26.**
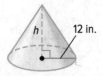
h   12 in.
Volume: 1,205.76 in³

**27.**
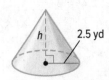
h   2.5 yd
Volume: 19.625 yd³

**28.** Four cones have the dimensions shown below.

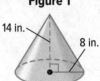

Figure 1

14 in.  8 in.

Figure 2

13 in.  9 in.

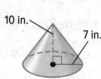

Figure 3

10 in.  7 in.

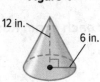

Figure 4

12 in.  6 in.

Sort the cones from least to greatest volume. Round to the nearest tenth.

| | Figure | Volume (in³) |
|---|---|---|
| Least | | |
| | | |
| | | |
| Greatest | | |

**29.** Refer to the cone shown at the right. Determine if each statement is true or false.

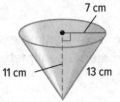

7 cm

11 cm    13 cm

a. The approximate area of the base is 153.9 square centimeters.   ☐ True  ☐ False

b. The approximate volume of the cone is 886.5 cubic centimeters.   ☐ True  ☐ False

c. The volume of a cylinder with the same height and radius would be 3 times the volume of the cone.   ☐ True  ☐ False

## CCSS **Common Core Spiral Review**

**Find the volume of each pyramid. Round to the nearest tenth if necessary.** 7.G.6

**30.**

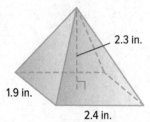

2.3 in.

1.9 in.

2.4 in.

**31.**

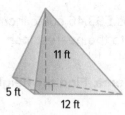

11 ft

5 ft

12 ft

**32.**

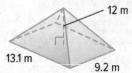

12 m

13.1 m

9.2 m

**33.**

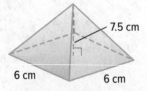

7.5 cm

6 cm    6 cm

# Volume of Spheres

## Vocabulary Start-Up

A **sphere** is a set of all points in space that are a given distance, known as the radius, from a given point, known as the center.

**Complete the graphic organizer.**

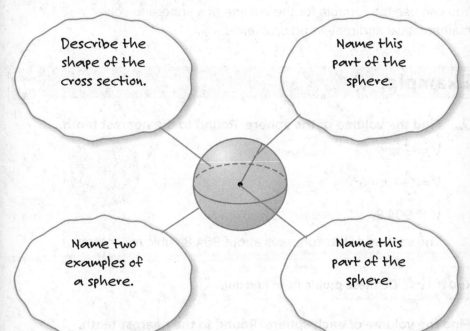

Describe the shape of the cross section.

Name this part of the sphere.

Name two examples of a sphere.

Name this part of the sphere.

### Essential Question

WHY are formulas important in math and science?

**Vocabulary**

sphere
hemisphere

### Common Core State Standards

**Content Standards**
8.G.9

**MP Mathematical Practices**
1, 3, 4

### Real-World Link

Brittani purchased a necklace that contained a round pearl with a diameter of 7.5 millimeters. What is the circumference of the largest circle around the outside of the pearl? Round to the nearest tenth. _____

**Which MP Mathematical Practices did you use?**
**Shade the circle(s) that applies.**

① Persevere with Problems
② Reason Abstractly
③ Construct an Argument
④ Model with Mathematics
⑤ Use Math Tools
⑥ Attend to Precision
⑦ Make Use of Structure
⑧ Use Repeated Reasoning

## Key Concept ▷ Volume of a Sphere

**Words** The volume $V$ of a sphere is four thirds the product of $\pi$ and the cube of the radius $r$.

**Symbols** $V = \frac{4}{3}\pi r^3$

**Model**

---

Work Zone

**Exact and Approximate**

Whenever you round or use 3.14 for $\pi$, you are finding the approximate value. An answer left in terms of $\pi$, such as $\frac{256}{3}\pi$, is an exact value.

You can use the formula for the volume of a sphere to solve mathematical and real-world problems.

Tutor

## Example

**1.** Find the volume of the sphere. Round to the nearest tenth.

$V = \frac{4}{3}\pi r^3$     Volume of a sphere

$V = \frac{4}{3} \cdot \pi \cdot 6^3$     Replace $r$ with 6.

$V \approx 904.8$     Simplify. Use a calculator.

The volume of the sphere is about 904.8 cubic millimeters.

---

Show your work.

**Got it?** Do these problems to find out.

Find the volume of each sphere. Round to the nearest tenth.

a.

b.

a. _____

b. _____

---

Real World

## Example

Tutor

**2.** A spherical stone in the courtyard of the National Museum of Costa Rica has a diameter of about 8 feet. Find the volume of the spherical stone. Round to the nearest tenth.

$V = \frac{4}{3}\pi r^3$     Volume of a sphere

$V = \frac{4}{3} \cdot \pi \cdot 4^3$     Replace $r$ with 4.

$V \approx 268.1$     Simplify. Use a calculator.

The volume of the spherical stone is about 268.1 cubic feet.

**Got it?** Do this problem to find out.

c. A dish contains a spherical scoop of vanilla ice cream with a radius of 1.2 inches. What is the volume of the ice cream?

c. _____

## Example

Tutor

**3.** **A volleyball has a diameter of 10 inches. A pump can inflate the ball at a rate of 325 cubic inches per minute. How long will it take to inflate the ball? Round to the nearest tenth.**

Find the volume of the ball. Then use a proportion.

$V = \frac{4}{3}\pi r^3$      Volume of a sphere

$V = \frac{4}{3} \cdot \pi \cdot 5^3$ or 523.6      Replace *r* with 5.

$\frac{325 \text{ in}^3}{1 \text{ min}} = \frac{523.6 \text{ in}^3}{x \text{ min}}$      Write the proportion.

$325x = 523.6$      Cross multiply.

$x = 1.6$      Simplify.

So, it will take about 1.6 minutes to inflate the ball.

**Got it?** Do this problem to find out.

d. _____

d. A snowball has a diameter of 6 centimeters. How long would it take the snowball to melt if it melts at a rate of 1.8 cubic centimeters per minute? Round to the nearest tenth.

# Volume of a Hemisphere

A circle separates a sphere into two congruent halves each called a **hemisphere**.

## Example

Tutor

**Hemisphere**

The volume of a hemisphere is $\frac{1}{2}$ the volume of a sphere.

**4.** **Find the volume of the hemisphere. Round to the nearest tenth.**

$V = \frac{1}{2}\left(\frac{4}{3}\pi r^3\right)$      Volume of a hemisphere

5 cm

$V = \frac{1}{2}\left(\frac{4}{3} \cdot \pi \cdot 5^3\right)$      Replace *r* with 5.

$V \approx 261.8$      Simplify. Use a calculator.

The volume of the hemisphere is about 261.8 cubic centimeters.

e. _____

f. _____

e.
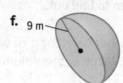
2 cm

f.  9 m

## Guided Practice

**Find the volume of each sphere. Round to the nearest tenth.** (Example 1)

**1.** _____

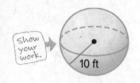
10 ft

Show your work.

**2.** _____

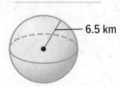
6.5 km

**3.** Sarah is blowing up spherical balloons for her brother's birthday party.
One of the balloons has a radius of 3 inches. Round to the nearest tenth.
(Examples 2 and 3)

a. What is the volume of the balloon? _____

b. Suppose Sarah can inflate the balloon at a rate of 200 cubic inches

per minute. How long will it take her to inflate the ballon? _____

**Find the volume of each hemisphere. Round to the nearest tenth.** (Example 4)

**4.** _____

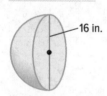
16 in.

**5.** _____

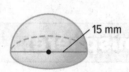
15 mm

**6.** Ⓔ **Building on the Essential Question** *True* or *false*?
The volume of a sphere is two-thirds the volume of a
cylinder with the same radius *r* and height of 2*r*.
Explain your reasoning.

_____

_____

_____

_____

### Rate Yourself!

How well do you understand
volume of spheres? Circle the
image that applies.

Clear    Somewhat    Not So
          Clear        Clear

For more help, go online to
access a Personal Tutor.

Tutor

# Extra Practice

Copy and Solve  For Exercises 16–36, show your work and answers on a separate piece of paper.

**Find the volume of each figure. Round to the nearest tenth.**

**16.**
11 cm

**17.**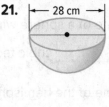
9 in.

**18.**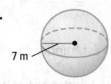
7 m

**19.**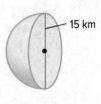
15 km

**20.**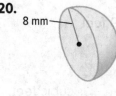
8 mm

**21.**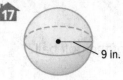
28 cm

**22.** Amber is purchasing a ring that contains a 7.5 millimeter diameter round pearl. Find the volume of the pearl to the nearest tenth.

**23.** Ms. Holliday is purchasing balloons for a party. Each spherical balloon is inflated with helium. How much helium is in the balloon if the balloon has a radius of 11 centimeters? Round to the nearest tenth.

**24.** (MP) **Persevere with Problems** The volume of a baseball is about 13.39 cubic inches. What is the diameter? Round to the nearest tenth.

**25.** A golf ball has a diameter of 42.67 millimeters and a mass of 45.93 grams. What is the number of grams per cubic millimeter of the material used to make the golf ball? Round to the nearest ten-thousandth.

**Find the volume of each composite solid. Round to the nearest tenth.**

**26.**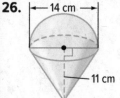
14 cm
11 cm

**27.**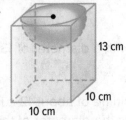
13 cm
10 cm
10 cm

**28.** The volume of a golf ball is about 41.63 cm³. Select the correct values to complete the formula below to find the radius of a golf ball.

| 2 | 6 |
|---|---|
| 3 | 9 |
| 4 | 41.63 |
|   | r |

$$\boxed{\phantom{xxxxx}} = \sqrt[\boxed{}]{\frac{\boxed{}}{\boxed{}\pi}\boxed{}^{\boxed{}}}$$

To the nearest hundredth, what is the radius of the golf ball? $\boxed{\phantom{xxxx}}$

**29.** Refer to the hemisphere shown. Fill in each box to make a complete statement. Round to the nearest tenth if necessary.

9 ft

**a.** The radius of the hemisphere is $\boxed{\phantom{xxxxx}}$ feet.

**b.** The volume of a sphere with a diameter of 9 feet is $\boxed{\phantom{xxxxx}}$ cubic feet.

**c.** The volume of the hemisphere is $\boxed{\phantom{xxxxx}}$ cubic feet.

## CCSS Common Core Spiral Review

**Find the circumference and area of each circle. Round to the nearest tenth.** 7.G.4

**30.**

3.6 in.

**31.**
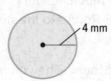
4 mm

**32.**
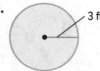
3 ft

**33.** ⊢— 6.2 m —⊣
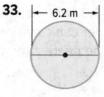

**34.** Find the area of a circle with a radius of 6 centimeters. Round to the nearest tenth. 7.G.4

**35.** Find the area of a circle with a diameter of 13.1 centimeters. Round to the nearest tenth. 7.G.4

**36.** An conical icicle has a volume of about 12 cubic inches. If the icicle has a height of 8 inches, what is the diameter of the icicle? 8.G.9

**Problem-Solving Investigation**

# Solve a Simpler Problem

**Content Standards**
8.G.9

MP **Mathematical Practices**
1, 4, 7

## Case #1  Spring Fling

Student government is decorating the school's gymnasium by hanging 100 of the containers shown from the ceiling. The cones have a radius of 3 inches and a height of 12 inches. The hemispheres have a radius of 3 inches.

*How much confetti will they need to fill one of the containers?*

### Understand  *What are the facts?*

Each container is made up of a cone and a hemisphere.

### Plan  *What is your strategy to solve this problem?*

Solve a simpler problem by separating the container into a cone and a hemisphere. Find the volume of each, and then add the two volumes together. Round your answers to the nearest tenth.

 = 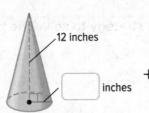   12 inches

□ inches +   3 inches

### Solve  *How can you apply the strategy?*

Volume of the cone = _____ in$^3$

Volume of the hemisphere = _____ in$^3$

Volume of one container = _____ in$^3$

So, one container will need _____ cubic inches of confetti.

### Check  *Does the answer make sense?*

The total volume is greater than each of the parts, so the answer is reasonable.

## Analyze the Strategy   Tutor

MP **Identify Structure** Suppose one bag contains 500 cubic inches of confetti. How many bags will Student Government need to fill 100 containers?

## Case #2 Carpenter's Riddle

Working separately, three carpenters can make three chairs in three days.

How many chairs can 7 carpenters working at the same rate make in 30 days?

## Understand

**Read the problem. What are you being asked to find?**

I need to find _____.

**Underline key words and values. What information do you know?**

_____ carpenters make _____ chairs in _____ days.

## Plan

**Choose a problem-solving strategy.**

I will use the _____ strategy.

## Solve

**Use your problem-solving strategy to solve the problem.**

Use the information provided.

If three carpenters can make three chairs in three days, then one carpenter can make _____ chair in _____ days.

If one carpenter can make _____ chair in _____ days, then one carpenter can make ☐ chairs in 30 days. If one carpenter makes ☐ chairs in 30 days, then 7 carpenters make

☐ × ☐ = ☐ chairs in 30 days.

So, _____.

## Check

**Use information from the problem to check your answer.**

_____

**Work with a small group to solve the following cases.**
**Show your work on a separate piece of paper.**

Collaborate

## Case #3  Storage

A 15-foot tall storage building is shown. Grain fills the storage
building to a height of 12 feet.

What is the volume of the space filled with grain?
Round your answer to the nearest tenth.

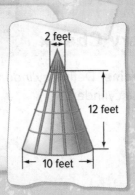

2 feet

12 feet

← 10 feet →

## Case #4  School Play

Four students can sew four costumes in two days.

How many costumes can ten students sew in twelve days?

## Case #5  Hidden Squares

Watch

Gina looked at the figure and decided there were
25 squares shown. Brandon told her there were many
more than that since there could be squares that
measure 1 × 1, 2 × 2, 3 × 3, 4 × 4, and 5 × 5.

How many squares of any size are in the
figure? (Hint: Count the number of squares
in a 2 × 2 and a 3 × 3 square. Then look for a
pattern.)

Use any
strategy!

## Case #6  Pizza

What is the largest number of
pieces that can be cut from
one pizza using five straight cuts?

3 cuts          4 cuts

## Vocabulary Check

1. **MP** **Be Precise** Define *cylinder*. What are the symbols used to find the volume of a cylinder? (Lesson 1)

_____

_____

**Fill in the blank.**

2. The volume of a cone is _____ the volume of a cylinder with the same base and height. (Lesson 2)

## Skills Check and Problem Solving

3. What is the volume of the cylinder shown at the right? Round to the nearest tenth. (Lesson 1) _____

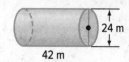

4. Find the height of a cone with a volume of 464.603 cubic feet and a diameter of 8 feet. (Lesson 2) _____

**Find the volume of each sphere. Round to the nearest tenth.** (Lesson 3)

5. _____

6. _____

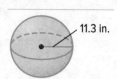

11.3 in.

7. _____

21 yd

8. **MP** **Reason Inductively** Refer to the cylinders shown. If a cone has a base and height congruent to Cylinder 1, which statement is true? (Lesson 2) _____

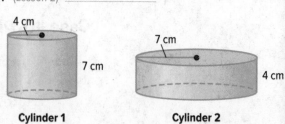

I  The volume of the cone is equal to the volume of Cylinder 1.

II  The volume of the cone is equal to the volume of Cylinder 2.

III  The cone has a greater volume than Cylinder 1.

IV  The cone has one third the volume of Cylinder 1.

# Inquiry Lab
## Surface Area of Cylinders

 **HOW can the surface area of a cylinder be determined?**

 **Content Standards**
Extension of 8.G.9

 **Mathematical Practices**
1, 3

Nyombi used a clean frozen yogurt container for a school project. The container is shaped like a cylindrical tub that has a diameter of 20 centimeters and a height of 30 centimeters. She wants to know how much paper she will need to cover the entire container.

## Hands-On Activity

**Nets** are two-dimensional patterns of three-dimensional figures. When you construct a net, you are decomposing the three-dimensional figure into separate shapes. You can use a net to find the area of each surface of a three-dimensional figure such as a cylinder.

**Step 1** Use an empty cylinder shaped container that has a lid.

What is the height of the container? _____

**Step 2** Take off the lid of the container and make 2 cuts as shown. Cut off the sides of the lid. Lay the lid, the curved side, and the bottom flat to form the net of the container. Sketch and label the parts of the net.

What are the shapes that make up the net of the

container? _____

**Step 3** Make a mark on the top of the lid. Place the mark at the top edge of the flattened curved side as shown. Roll the lid along the edge of the side until it completes one rotation.

Where does the lid stop? _____

How does the length of the curved side compare to the distance

around the top? _____

Find the area of each shape.

Top _____ Bottom _____ Side _____

Find the sum of these areas. _____

## Investigate

Collaborate

Work with a partner. Draw the net and label the parts of the cylinder and the measurements. Then complete the table to find the *total surface areas* for Exercises 1 and 2. Round to the nearest tenth.

**1.**

Show your work.

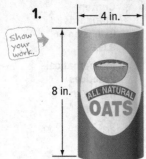

← 4 in. →

8 in.

**2.**

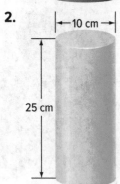

← 10 cm →

25 cm

| | Area of top ($\pi r^2$) | Area of bottom ($\pi r^2$) | Curved Area | Total Surface Area |
|---|---|---|---|---|
| **3.** | | | | |
| **4.** | | | | |

## Analyze and Reflect

Collaborate

**5.** What is the total surface area of the container described at the beginning of the lesson? Round to the nearest tenth. _____

## Create

On Your Own

**6.** **MP** **Reason Inductively** Describe how to find the area of the curved surface of a cylinder. _____

_____

_____

**7.** **Inquiry** How can the surface area of a cylinder be determined?

_____

_____

# Surface Area of Cylinders

 **Real-World Link**

**Bakery** The Shiny Bright bakery is making a cake for Maria's quinceañera. The cake will be in the shape of a cylinder with a height of 4 inches and a diameter of 14 inches.

1. What are the shapes that make up the net of the cake? Sketch the net in the space provided.

   _____

2. How is the length of the rectangle related to the circles that form the top and bottom of the cake?

   _____

   _____

3. Find the area of each part of the cake. Round to the nearest whole number.

   Top: [   ] in²    Bottom: [   ] in²    Side: [   ] in²

4. Add the values from Exercise 3. What is the total surface area of the cake? [   ] in²

 **Essential Question**

WHY are formulas important in math and science?

 Vocab **Vocabulary**

lateral area
total surface area

 **Common Core State Standards**

**Content Standards**
Extension of 8.G.9

MP **Mathematical Practices**
1, 3, 4

Which MP **Mathematical Practices** did you use?
Shade the circle(s) that applies.

① Persevere with Problems    ⑤ Use Math Tools

② Reason Abstractly    ⑥ Attend to Precision

③ Construct an Argument    ⑦ Make Use of Structure

④ Model with Mathematics    ⑧ Use Repeated Reasoning

# Surface Area of a Cylinder

### Lateral Area

**Words**  The lateral area *L.A.* of a cylinder with height *h* and radius *r* is the circumference of the base times the height.

**Symbols**  $L.A. = 2\pi rh$

### Total Surface Area

**Words**  The surface area *S.A.* of a cylinder with height *h* and radius *r* is the lateral area plus the area of the two circular bases.

**Symbols**  $S.A. = L.A. + 2\pi r^2$ or $S.A. = 2\pi rh + 2\pi r^2$

**Model**

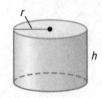

area of base $= \pi r^2$

You can find the surface area of a cylinder using a net.

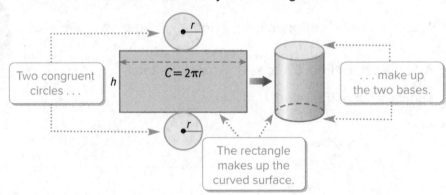

Two congruent circles . . .

$C = 2\pi r$

. . . make up the two bases.

The rectangle makes up the curved surface.

In the diagram above, the length of the rectangle is the same as the circumference of the circle, $2\pi r$. Also, the width of the rectangle is the same as the height of the cylinder.

The **lateral area** of a three-dimensional figure is the surface area of the figure, excluding the area of the base(s). So, the lateral area of a cylinder is the area of curved surface.

The **total surface area** of a three-dimensional figure is the sum of the areas of all its surfaces.

## Example

Tutor

**1.** Find the surface area of the cylinder.
Round to the nearest tenth.

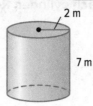
2 m
7 m

$S.A. = 2\pi rh + 2\pi r^2$      Surface area of a cylinder

$S.A. = 2\pi(2)(7) + 2\pi(2)^2$      Replace $r$ with 2 and $h$ with 7.

$S.A. \approx 113.1$      Simplify.

The surface area is about 113.1 square meters.

**Got it?** Do these problems to find out.

*Show your work.*

Find the surface area of each cylinder. Round to the nearest tenth.

**a.**
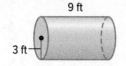
9 ft
3 ft

**b.**

←14 cm→
10 cm

a. _____

b. _____

## Example

Tutor

**2.** A circular fence that is 2 feet high is to be built around the outside of a carousel. The distance from the center of the carousel to the edge of the fence will be 35 feet. What is the area of the fencing material that is needed to make the fence around the carousel?

You need to find the lateral area. The radius of the circular fence is 35 feet. The height is 2 feet.

$L.A. = 2\pi rh$      Lateral area of a cylinder

$L.A. = 2\pi(35)(2)$      Replace $r$ with 35 and $h$ with 2.

$L.A. \approx 439.8$      Simplify.

So, about 439.8 square feet of material is needed to make the fence.

**Got it?** Do these problems to find out.

**c.** Find the area of the label of a can of tuna with a radius of 5.1 centimeters and a height of 2.9 centimeters. Round to the nearest tenth.

c. _____

**d.** Find the total surface area of a cylindrical candle with a diameter of 4 inches and a height of 2.5 inches. Round to the nearest tenth.

d. _____

**Find the total surface area of each cylinder. Round to the nearest tenth.** (Example 1)

**1.** _____

Show your work.

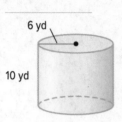
6 yd

10 yd

**2.** _____

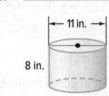
← 11 in. →

8 in.

**3.** _____

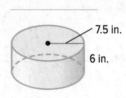
7.5 in.

6 in.

**4.** _____

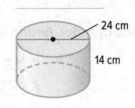
24 cm

14 cm

**5.** Find the total surface area of a water tank with a height of 10 meters and a diameter of 10 meters. Round to the nearest tenth. (Example 1) _____

**Find the lateral area of each cylinder. Round to the nearest tenth.** (Example 2)

**6.** _____

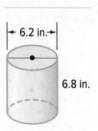
← 6.2 in. →

6.8 in.

**7.** _____

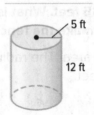
5 ft

12 ft

**8.** Find the area of the label of a cylindrical potato chip container with a radius of 3.1 inches and a height of 9.2 inches. Round to the nearest tenth. (Example 2)

_____

_____

**9.** ⓔ **Building on the Essential Question** How is a calculation affected if you round π to 3.14 or use the π key on your calculator? Explain.

_____

_____

_____

**Rate Yourself!**

Are you ready to move on? Shade the section that applies.

I have a few questions. | I'm ready to move on.

I have a lot of questions.

For more help, go online to access a Personal Tutor.

Tutor

FOLDABLES Time to update your Foldable!

Name _____ My Homework _____

# Independent Practice

Go online for Step-by-Step Solutions

**Find the total surface area of each cylinder. Round to the nearest tenth.**
(Example 1)

**1.** _____

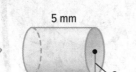

5 mm

2 mm

**2.** _____

12.5 m

9 m

**3.** _____

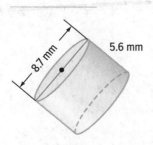

8.7 mm

5.6 mm

**4.** _____

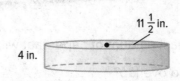

11 $\frac{1}{2}$ in.

4 in.

**5.** A cylindrical candle has a diameter of 4 inches and a height of 7 inches. To the nearest tenth, what is the total surface area of the

candle? (Example 1) _____

**6.** Find the total surface area of an unsharpened cylindrical pencil that has a radius of 0.5 centimeter and a height of 19 centimeters. Round to the

nearest tenth. (Example 1) _____

**Find the lateral area of each cylinder. Round to the nearest tenth.** (Example 2)

**7.** _____

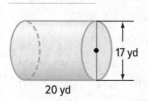

17 yd

20 yd

**8.** _____

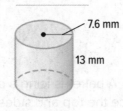

7.6 mm

13 mm

**9** Find the lateral area of a cylindrical copper pipe that has a diameter of 6.4 inches and a height of 12 inches. Round to the nearest tenth.

(Example 2) _____

10. **MP Model with Mathematics** Refer to the graphic novel frame below.

I think we should wrap these in gift paper before we put them in the bags.

You can review the dimensions of the candles in Lesson 1.

a. What is the least amount of paper that will be needed to wrap one candle with no overlap? _____

b. How many square feet of wrapping paper will be needed to wrap all 70 candles? _____

## H.O.T. Problems Higher Order Thinking

11. **MP Persevere with Problems** If the height of a cylinder is doubled, will its surface area also double? Explain your reasoning.

_____

_____

12. **MP Reason Inductively** Which has a greater surface area: a cylinder with radius 6 centimeters and height 3 centimeters or a cylinder with radius 3 centimeters and height 6 centimeters? Explain your reasoning.

_____

_____

_____

13. **MP Reason Inductively** A baker is icing a cylindrical cake with radius $r$ and height $h$. The baker will ice the top and sides of the cake. Write an equation giving the total area $A$ that the baker will ice. Explain why your equation is not the same as the formula for the total surface area of a cylinder.

_____

_____

_____

# Extra Practice

**Copy and Solve** For Exercises 14–27, show your work and answers on a separate piece of paper.

Find the lateral area and the total surface area of each cylinder. Round to the nearest tenth.

14.
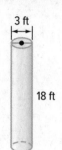
3 ft

18 ft

15.
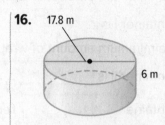
22 cm

16 cm

16. 17.8 m
6 m

17. A lamp shade is in the shape of a cylinder with a height of 18 inches and a radius of $6\frac{3}{4}$ inches. Fabric will cover the lateral area of the lamp shade. Find the area of the fabric needed. Round to the nearest tenth.

**MP Use Math Tools** Estimate the surface area of each cylinder.

18.
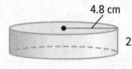
4.8 cm

2.2 cm

19.
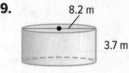
8.2 m

3.7 m

20. 12.8 ft
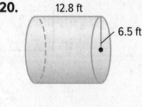
6.5 ft

21. The mail tube shown is made of cardboard and has plastic end caps. Approximately what percent of the surface area of the mail tube is cardboard?

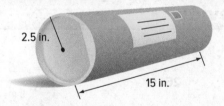
2.5 in.

15 in.

22. **MP Persevere with Problems** A hot cocoa canister is a cylinder with a height of 24.5 centimeters and a diameter of 13 centimeters.

a. What is the lateral area of the hot cocoa canister to the nearest tenth?

b. How does the lateral area change if the height is divided by 2?

13 cm

24.5 cm

Hot Cocoa Mix

23. Maria is comparing how much wrapping paper it would take to wrap the containers below.

    Complete each statement with the correct answers. Round answers to the nearest square inch.

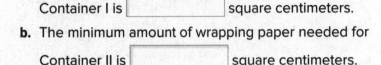

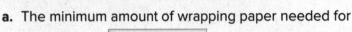

**Container I**   **Container II**

   a. The minimum amount of wrapping paper needed for Container I is [ ] square centimeters.

   b. The minimum amount of wrapping paper needed for Container II is [ ] square centimeters.

   c. Container [ ] requires [ ] more square centimeters of wrapping paper than Container [ ].

24. Stacey has a cylindrical paper clip holder with a diameter of 2 inches and a height of 1.5 inches. Label the net of the cylinder below with the correct dimensions.

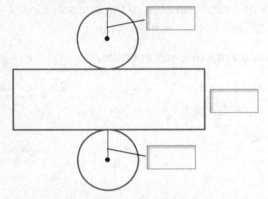

    To the nearest tenth, what is the surface area of the paper clip holder? [ ]

**Find the surface area of each prism.** 7.G.6

25.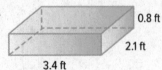
    0.8 ft
    2.1 ft
    3.4 ft

26.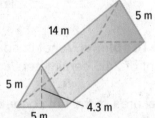
    5 m
    14 m
    5 m
    5 m
    4.3 m

27. Mrs. Jones is filling cone-shaped treat bags with candy. Each bag has a height of 6 inches and a radius of 0.75 inch. What is the volume of each bag? Round to the nearest tenth. 8.G.9

# Inquiry Lab
## Nets of Cones

 **Inquiry** HOW can the surface area of a cone be found?

Corinne is making party hats as decorations for a party. They are in the shape of a cone and will be covered with tissue paper. What is the surface area of the cone that will be covered with tissue paper?

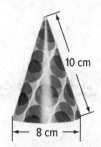

10 cm

8 cm

 **Content Standards** Extension of 8.G.9

**MP** **Mathematical Practices** 1, 3

## Hands-On Activity 1

In this Activity, you will construct a net of the cone for the party hats. The radius of the base is 4 centimeters. The *slant height* of the cone is 10 centimeters.

**Step 1** On a separate sheet of paper, use a compass to draw two circles slightly touching, one with a radius of 10 centimeters and one with a radius of 4 centimeters.

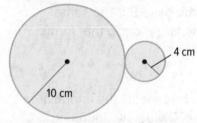

4 cm

10 cm

**Step 2** You need to find the portion of the circumference from the larger circle that will wrap around the outside of the smaller circle to make the cone. Use the proportion shown to find the central angle measure that represents the portion of the large circle you will use.

$$\frac{\text{circumference of smaller circle}}{\text{circumference of larger circle}} = \frac{\text{portion (in degrees) that is unknown}}{\text{total number of degrees in a circle}}$$

Write and solve the proportion. Round to the nearest whole number.

$$\frac{\boxed{\phantom{xx}}}{\boxed{\phantom{xx}}} = \frac{x}{360}$$

$$x \approx \boxed{\phantom{xx}}$$

So, you will need $\boxed{\phantom{xx}}$ degrees of the larger circle.

**Step 3** Use a protractor to draw the central angle in your larger circle to create the net of the cone.

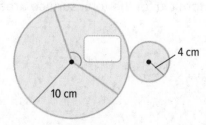

4 cm

10 cm

**Step 4**   The net of the cone is the portion of both circles that are shown by the solid lines. Cut out the net and make the cone.

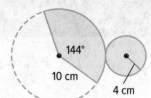

10 cm

4 cm

# Hands-On Activity 2

The net shows that the surface area of a cone with slant height $\ell$ and radius $r$ is the sum of its base $B$ and its lateral area $L.A.$ The base $B$ is a circle. The lateral area $L.A.$ is *part* of a larger circle.

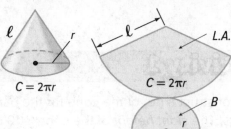

The circumference of the base $B$ is the same length as the part of the larger circle that forms the lateral area of the cone.

**Step 1**   The figure represents the lateral area of the cone. Divide the figure into 6 equal sections. The first one is done for you.

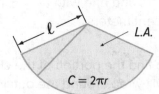

**Step 2**   The parallelogram shows the 6 rearranged sections. Write an expression that represents the length of the parallelogram.

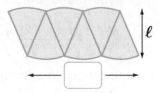

**Step 3**   Use the expression from Step 2 to write a formula for the area of the parallelogram, which is the lateral surface area of the cone. _____

**Step 4**   Write a formula for the total surface area of the cone. _____

## Investigate

Collaborate

**Work with a partner. Draw the net of a cone in the space provided with each of the following dimensions.**

1. base radius: 1 inch

   slant height: 1.5 inches

   The angle measure needed to create the cone is _____°.

2. base radius: 2 centimeters

   slant height: 4 centimeters

   The angle measure needed to create the cone is _____°.

## Analyze and Reflect

**Collaborate**

Work with a partner. Use the formula from Activity 2 to find the total surface area of each of the following cones given the radius of the base and the slant height. Round the measure of the central angle to the nearest whole number. Round the surface area to the nearest tenth.

| | radius of base (r) | slant height (ℓ) | measure of central angle (°) | surface area ($\pi r\ell + \pi r^2$) |
|---|---|---|---|---|
| **3.** | 2 ft | 5 ft | | |
| **4.** | 5 in. | 15 in. | | |
| **5.** | 3 cm | 20 cm | | |

**6.** Refer to Activity 1. What is the lateral area of the party hat that Corinne is covering with tissue paper? Round to the nearest tenth.

## Create

**On Your Own**

**7.** **MP Make a Conjecture** Suppose the radius of the base of a cone is increased while the slant height stays the same. Make a conjecture about how the lateral surface area is affected.

**8.** **MP Make a Conjecture** Suppose a cone's slant height is decreased. Make a conjecture about which is affected more: the base or the lateral area. Justify your response.

**9.** **Inquiry** HOW can the surface area of a cone be found?

# Surface Area of Cones

## Vocabulary Start-Up

Recall that a cone is a three-dimensional figure with one circular base. A curved surface connects the base and vertex.

**Complete the graphic organizer.**

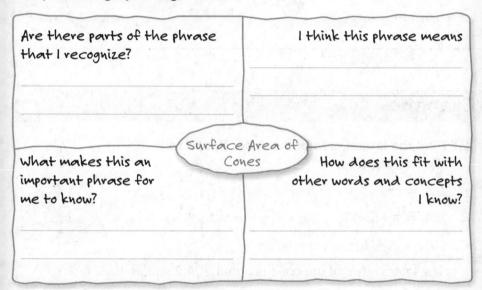

Are there parts of the phrase that I recognize?

I think this phrase means

Surface Area of Cones

What makes this an important phrase for me to know?

How does this fit with other words and concepts I know?

**Essential Question**

WHY are formulas important in math and science?

**Common Core State Standards**

**Content Standards**
Extension of 8.G.9

**MP Mathematical Practices**
1, 2, 3, 7

## Real-World Link

Bobbie is making waffle cones from scratch. Use the Pythagorean Theorem to find the slant height $\ell$ of the cone if the radius is 2 inches and the height is 6 inches.

Round to the nearest tenth. ☐ in.

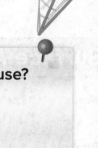

2 in.

6 in.

$\ell$

Which **MP Mathematical Practices** did you use?
Shade the circle(s) that applies.

① Persevere with Problems

⑤ Use Math Tools

② Reason Abstractly

⑥ Attend to Precision

③ Construct an Argument

⑦ Make Use of Structure

④ Model with Mathematics

⑧ Use Repeated Reasoning

## Key Concept ⟩ Lateral Area of a Cone

**Words**   The lateral area *L.A.* of a cone is π times the radius times the slant height ℓ.

**Symbols**   $L.A. = \pi r \ell$

**Model**

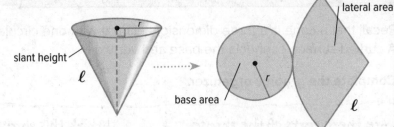

slant height  ℓ     base area     lateral area     ℓ     r

### Work Zone

**Lateral Area of a Cone**

The lateral area of a cone is one-half the circumference of the base times the slant height.

$L.A. = \frac{1}{2}(2\pi r)l$

$L.A. = \pi r l$

## Example

Tutor

1. **Find the lateral area of the cone. Round to the nearest tenth.**

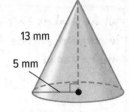
13 mm
5 mm

$L.A. = \pi r \ell$       Lateral area of a cone

$L.A. = \pi \cdot 5 \cdot 13$       Replace *r* with 5 and ℓ with 13.

$L.A. \approx 204.2$       Simplify.

The lateral area of the cone is about 204.2 square millimeters.

**Got it?** Do these problems to find out.

Show your work.

a. Find the lateral area of a cone with a radius of 4 inches and a slant height of 9.5 inches. Round to the nearest tenth.

b. Find the lateral area of a cone with a diameter of 16 centimeters and a slant height of 10 centimeters. Round to the nearest tenth.

a. _____

b. _____

## Key Concept ⟩ Surface Area of a Cone

**Words**   The surface area *S.A.* of a cone with slant height ℓ and radius *r* is the lateral area plus the area of the base.

**Symbols**   $S.A. = L.A. + \pi r^2$ or $S.A. = \pi r \ell + \pi r^2$

You can find the surface area of a cone using a net. The surface area of a cone is the sum of its lateral area and the area of its base.

**Model of Cone**

**Net of Cone**

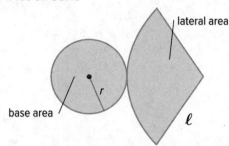

lateral area

base area

# Example

2. **Find the surface area of the cone. Round to the nearest tenth.**

6.2 inches

6 inches

$S.A. = \pi r\ell + \pi r^2$  　Surface area of a cone

$S.A. = \pi \cdot 6 \cdot 6.2 + \pi \cdot 6^2$  　Replace $r$ with 6 and $\ell$ with 6.2.

$S.A. \approx 230.0$  　Simplify.

The surface area of the cone is about 230.0 square inches.

**Got it?** Do this problem to find out.

c. Find the surface area of the cone. Round to the nearest tenth.

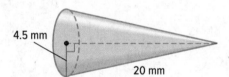

4.5 mm

20 mm

*Show your work.*

c. _____

# Real World Example

3. **A tepee has a radius of 5 feet and a slant height of 12 feet. Find the lateral area of the tepee. Round to the nearest tenth.**

$L.A. = \pi r\ell$  　Lateral area of a cone

$L.A. = \pi \cdot 5 \cdot 12$  　Replace $r$ with 5 and $\ell$ with 12.

$L.A. \approx 188.5$  　Simplify.

The lateral area of the tepee is about 188.5 square feet.

**Got it?** Do this problem to find out.

d. Ralph bought party hats that were in the shape of a cone. Each hat has a diameter of 8 inches and a slant height of 11 inches. Find the lateral area of one hat. Round to the nearest tenth.

d. _____

**Find the lateral area of each cone. Round to the nearest tenth.** (Example 1)

1. _____

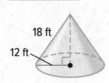

18 ft

12 ft

2. _____

6 m

30 m

3. _____

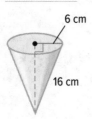

6 cm

16 cm

4. _____

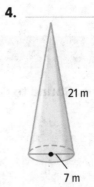

21 m

7 m

**Find the surface area of each cone. Round to the nearest tenth.** (Example 2)

5. _____

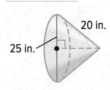

20 in.

25 in.

6. _____

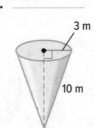

3 m

10 m

7. A local ice cream shop sells waffle cones dipped in chocolate. The waffle cone has a diameter $2\frac{5}{8}$ inches and a slant height of 6 inches. Find the lateral area of the waffle cone. Round to the nearest tenth. (Example 3)

_____

8. @ **Building on the Essential Question** How does the volume of a three-dimensional figure differ from its surface area?

_____

_____

_____

Name _____ My Homework _____

# Independent Practice

Go online for Step-by-Step Solutions

**Find the lateral area of each cone. Round to the nearest tenth.** (Example 1)

**1.** _____

10.2 in.
8.4 in.

**2.** _____

35 mm
18 mm

**3.** _____

25 m

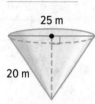

20 m

**Find the surface area of each cone. Round to the nearest tenth.** (Example 2)

**4.** _____

17 in.   22 in.

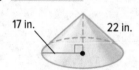

**5.** _____

9 cm

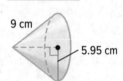

5.95 cm

**6.** _____

16 m

7 m

**7** A snow cone has a diameter of 1.9 inches and a slant height of 4.5 inches. What is the lateral area of the snow cone? Round to the nearest tenth. (Example 3)

_____

**8.** An active conical volcano has a radius of about 2.5 kilometers and slant height of about 9.6 kilometers. What is the lateral area of the volcano? Round to the nearest

tenth. (Example 3) _____

**9.** The lateral area of a cone with a diameter of 15 millimeters is about 333.5 square millimeters.
   **a.** Find the surface area of the cone. Round to the nearest tenth.

   _____

   **b.** What is the slant height of the cone? Round to the nearest tenth.

   _____

10. **MP Identify Structure** Match the figure with its correct volume or surface area formula.

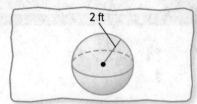

2 ft

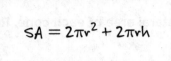

$$SA = 2\pi r^2 + 2\pi rh$$

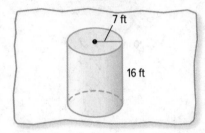

7 ft

16 ft

$$SA = \pi r l + \pi r^2$$

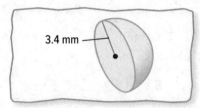

3.4 mm

$$V = \frac{4}{3}\pi r^3$$

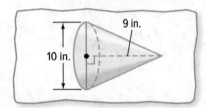

9 in.

10 in.

$$V = \frac{2}{3}\pi r^3$$

## H.O.T. Problems  Higher Order Thinking

11. **MP Find The Error**  Enrique is finding the surface area of a traffic cone. The traffic cone has a diameter of 10 inches and a height of 12 inches. Find his mistake and correct it.

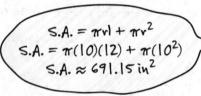

S.A. $= \pi r l + \pi r^2$
S.A. $= \pi(10)(12) + \pi(10^2)$
S.A. $\approx 691.15$ in$^2$

12. **MP Persevere with Problems**  Draw a cone with a surface area that is between 50 and 100 square units.

13. **MP Justify Conclusions**  Which has a greater surface area: a square pyramid with a base of $x$ units and a slant height of $\ell$ units or a cone with a diameter of $x$ units and a slant height of $\ell$ units? Explain your reasoning.

Name _____ My Homework _____

# Extra Practice

Copy and Solve For Exercises 14–35, show your work and answers on a separate piece of paper.

**Find the lateral area of each cone. Round to the nearest tenth.**

**14.** 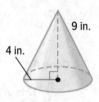 18 in. 15 in.

**15** 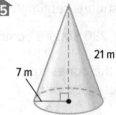 21 m 7 m

**16.** 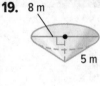 14 m 16.4 m

**17.** 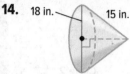 9 in. 4 in.

**18.**  11.7 mm 2 mm

**19.** 8 m 5 m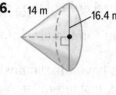

**20.** Find the lateral area of a cone with a radius of 3.5 millimeters and a slant height of 8 millimeters. Round to the nearest tenth.

**21.** Find the lateral area of a cone with a radius of 9 inches and a slant height of 16 inches. Round to the nearest tenth.

**Find the surface area of each cone. Round to the nearest tenth.**

**22.** 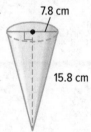 7.8 cm 15.8 cm

**23.** 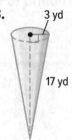 3 yd 17 yd

**24.** Find the surface area of a cone with a diameter of 20 millimeters and a slant height of 42 millimeters. Round to the nearest tenth.

**25.** Find the surface area of a cone with a radius of 5.1 feet and a slant height of 17 feet. Round to the nearest tenth.

**26.** (MP) **Reason Abstractly** A conical hat has a radius of 7 inches and a height of 14 inches. Find the slant height of the hat. Then find the lateral area. Round to the nearest tenth.

Lesson 5 Surface Area of Cones **637**

**27.** A cone has the radius and height shown.
Which of the following statements are true?
Select all that apply.

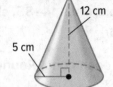

☐ The slant height of the cone is 13 cm.

☐ The lateral area of the cone is about 204 square centimeters.

☐ The total surface area of the cone is about 236 square centimeters.

**28.** Four cones have the dimensions shown. Sort the cones from least to greatest lateral areas. Round to the nearest tenth.

| | Cone | Lateral Area (m²) |
|---|---|---|
| Least | | |
| | | |
| | | |
| Greatest | | |

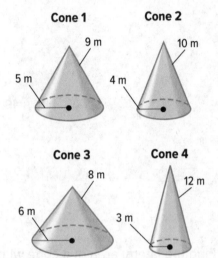

Cone 1 · Cone 2 · Cone 3 · Cone 4

---

## CCSS **Common Core Spiral Review**

**Find the surface area of each cylinder. Round to the nearest tenth.** 8.G.9

**29.**  3 ft / 5 ft

**30.** ←8 yd→ / 10 yd

**31.** 6 cm / 16 cm

**32.** diameter: 10 meters
height: 24 meters

**33.** radius: 12 feet
height: 9 feet

**34.** Find the volume of a cylinder with a radius of 2 inches and a height of 25 inches. Round to the nearest tenth. 8.G.9

**35.** Find the volume of a cone with a diameter of 16 feet and a height of 26 feet. Round the nearest tenth. 8.G.9

# Inquiry Lab
## Changes in Scale

 **Inquiry**   HOW does multiplying the dimensions of a three-dimensional figure by a scale factor affect its volume and surface area?

 **Content Standards**
8.G.9

**MP Mathematical Practices**
1, 3

A cake decorator is making a large center cake surrounded by small cakes. The large cake has a diameter of 24 inches and is 8 inches tall. The small cakes' dimensions will be half the dimensions of the large cake. Use the Activity to determine how the volume and surface area of the large cake compare to the volume and surface area of one of the small cakes.

## Hands-On Activity

In this Activity, you will determine how changes in the dimensions of a cylinder affect the volume and surface area.

**Step 1**   Complete the table with the measurements of the cakes.

| | Diameter (in.) | Radius (in.) | Height (in.) |
|---|---|---|---|
| **Large Cake** | | | |
| **Small Cake** | | | |

**Step 2**   Find the volume of each of the cakes. Round to the nearest whole number.

Volume of large cake = _____

Volume of small cake = _____

The volume of the large cake is about ⬜ times the volume of a small cake.

**Step 3**   Find the surface area of each cake. Round to the nearest whole number.

Surface area of large cake = _____

Surface area of small cake = _____

The surface area of the large cake is ⬜ times the surface area of a small cake.

## Investigate

**Work with a partner. Round your answers to the nearest whole number.**

1. The radius of a cylinder is 25 centimeters and the height is 35 centimeters. The dimensions of a similar cylinder are one-fifth of the original cylinder. Complete the chart.

Show your work.

|  | Volume | Surface Area |
|---|---|---|
| original cylinder |  |  |
| similar cylinder |  |  |
| similar / original |  |  |

2. The diameter of a cone is 6 inches, the height is 4 inches, and the slant height is 5 inches. The dimensions of a similar cone are three times that of the original cone.

|  | Volume | Surface Area |
|---|---|---|
| original cone |  |  |
| similar cone |  |  |
| similar / original |  |  |

## Analyze and Reflect

**Work with a partner.**

3. Find the volume and surface area for the cylinders shown in the table. Use 3.14 for π.

4. What happens to the volume of a cylinder when the radius and height are multiplied by two? by three?

| Cylinder | Radius and Height | Volume | Surface Area |
|---|---|---|---|
| A | 1 |  |  |
| B | 2 |  |  |
| C | 3 |  |  |

5. What happens to the surface area of a cylinder when the radius and height are multiplied by two? by three? _____

## Create

On Your Own

6. Suppose the dimensions of a rectangular prism are doubled. The volume of the new prism is 800 cubic units. What are the possible dimensions of the original prism? _____

7. **inquiry** HOW does multiplying the dimensions of a three-dimensional figure by a scale factor affect its volume and surface area?

_____

_____

# Changes in Dimensions

## Real-World Link

 Watch

**Monuments** Stephen is creating a model of the Washington Monument for history class. The model will be $\frac{1}{100}$ of the monument's actual size.

slant height $\ell$

The square pyramid that sits atop the monument's obelisk shape has a slant height of about 57.6 feet. Each side of the pyramid's base is about 34 feet.

1. What is the area of one of the triangular faces of the actual pyramid? _____

2. What is the slant height of the pyramid on the model Stephen is creating? _____

3. What is the length of one side of the base of the pyramid on the model? _____

4. What is the area of one of the triangular faces of the model pyramid? _____

5. Write a ratio comparing the area of the triangular side of the model to the actual monument.

_____

6.  **Make a Conjecture** Write a sentence about the surface area of the model pyramid compared with the actual pyramid.

_____

 **Essential Question**

WHY are formulas important in math and science?

Vocab
 **Vocabulary**

similar solids

CCSS **Common Core State Standards**

Content Standards
8.G.9

MP **Mathematical Practices**
1, 3, 4

Which MP **Mathematical Practices** did you use?
Shade the circle(s) that applies.

① Persevere with Problems
② Reason Abstractly
③ Construct an Argument
④ Model with Mathematics
⑤ Use Math Tools
⑥ Attend to Precision
⑦ Make Use of Structure
⑧ Use Repeated Reasoning

# Surface Area of Similar Solids

If Solid X is similar to Solid Y by a scale factor, then the surface area of X is equal to the surface area of Y times the *square* of the scale factor.

Cubes are **similar solids** because they have the same shape and their corresponding linear measures are proportional.

The cubes at the right are similar. The ratio of their corresponding edge lengths is $\frac{8}{4}$ or 2. The scale factor is 2. How are their surface areas related?

 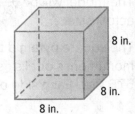

4 in.
4 in.
4 in.

8 in.
8 in.
8 in.

**S.A. of Small Cube**

S.A. = 6(4 · 4)

There are 6 faces.

**S.A. of Large Cube**

S.A. = 6(2 · 4)(2 · 4)

= 2 · 2(6)(4 · 4)

= $2^2$(6)(4 · 4)

To find the surface area of the large cube, multiply the surface area of the small cube by the *square* of the scale factor, $2^2$ or 4. This relationship is true for any similar solids.

## Example

 Tutor

**1.** The surface area of a rectangular prism is 78 square centimeters. What is the surface area of a similar prism with dimensions that are 3 times as great as the dimensions of the original prism?

S.A. = 78 × $3^2$    Multiply by the square of the scale factor.

S.A. = 78 × 9    Square 3.

S.A. = 702 cm$^2$    Simplify.

 Show your work.

**Got it?** Do these problems to find out.

a. _____

**a.** The surface area of a triangular prism is 34 square inches. What is the surface area of a similar prism with dimensions that are 3 times as great as the original prism?

b. _____

**b.** The world's largest box of raisins has a surface area of 352 square feet. If the dimensions of a similar box are smaller than the largest box by a scale factor of $\frac{1}{48}$, what is its surface area?

# Volume of Similar Solids

If Solid X is similar to Solid Y by a scale factor, then the volume of X is equal to the volume of Y times the *cube* of the scale factor.

Refer to the cubes below.

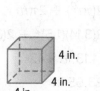
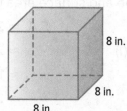

**Volume of Small Cube**
$V = 4 \cdot 4 \cdot 4$

**Volume of Large Cube**
$V = (2 \cdot 4)(2 \cdot 4)(2 \cdot 4)$
$= 2 \cdot 2 \cdot 2(4 \cdot 4 \cdot 4)$
$= 2^3(4 \cdot 4 \cdot 4)$

The volumes of similar solids are related by the *cube* of the scale factor.

# Example

Tutor

**2.** A triangular prism has a volume of 432 cubic yards. If the dimensions of the prism are reduced to one third of the original dimensions, what is the volume of the new prism?

$V = 432 \times \left(\frac{1}{3}\right)^3$  Multiply by the cube of the scale factor.

$V = 432 \times \frac{1}{27}$  Cube $\frac{1}{3}$.

$V = 16 \text{ yd}^3$  Simplify.

The volume of the new prism is 16 cubic yards.

**Got it?** Do these problems to find out.

Show your work.

**c.** A square pyramid has a volume of 512 cubic centimeters. What is the volume of a square pyramid with dimensions one-fourth of the original?

c. _____

**d.** A cylinder has a volume of 432 cubic meters. What is the volume of a cylinder with dimensions one-third of the original?

d. _____

 **Example**

**3.** The measurements for a standard hockey puck are shown at the right. The giant hockey puck at the left has dimensions that are 40 times the dimensions of a standard puck. Find the volume and surface area of the giant puck. Use 3.14 for π.

1.5 in.

1 in.

Find the volume and surface area of the standard puck first.

$V = \pi r^2 h$          $S.A. = 2(\pi r^2) + 2\pi rh$

$\approx (3.14)(1.5)^2(1)$      $\approx 2(3.14)(1.5)^2 + 2(3.14)(1.5)(1)$

$\approx 7.065\ \text{in}^3$        $\approx 14.13 + 9.42$

                    $\approx 23.55\ \text{in}^2$

**STOP and Reflect**

What happens to the surface area of a cylinder if its radius and its height are doubled?

Find the volume and surface area of the giant puck using the computations for the standard puck and the scale factor.

$V = V(40)^3$          $S.A. = S.A.(40)^2$

$= (7.065)(40)^3$     $= (23.55)(40)^2$

$= 452,160\ \text{in}^3$      $= 37,680\ \text{in}^2$

The giant hockey puck has a volume of about 452,160 cubic inches and a surface area of about 37,680 square inches.

## Guided Practice

**1.** The surface area of a rectangular prism is 35 square inches. What is the surface area of a similar solid with dimensions that have been enlarged by

 a scale factor of 7? (Example 1) _____

**2.** The volume of a cylinder is about 425 cubic centimeters. What is the volume, to the nearest tenth, of a similar solid with dimensions that are

smaller by a scale factor of $\frac{1}{3}$? (Example 2) _____

**3.** A sink with a sliding lid in Josh's art studio measures 16 inches by 15 inches by 6 inches. A second sink used just for paintbrushes has a similar shape and is smaller

by a scale factor of $\frac{1}{2}$. Find the volume and surface area

of the second sink. (Example 3) _____

**Rate Yourself!**

How confident are you about changes in dimensions? Check the box that applies.

**4.**  **Building on the Essential Question** How is the volume of a prism affected when its dimensions are tripled?

_____

For more help, go online to access a Personal Tutor.

# Independent Practice

Go online for Step-by-Step Solutions

1. The surface area of a rectangular prism is 95 square centimeters. What is the surface area of a similar prism with dimensions that are 4 times as great as the original prism? (Example 1) _____

2. The surface area of a pyramid is 57.8 square inches. What is the surface area of a similar pyramid with dimensions that are 2 times as great as the original prism? (Example 1) _____

3. A cereal box has a surface area of 280 square inches. What is the surface area of a similar box that is larger by a scale factor of 1.4? (Example 1) _____

4. A glass display box has a surface area of 378 square inches. How many square inches of glass are used to create a glass display box with dimensions that are one-half those of the original? (Example 1) _____

5. A cone has a volume of 9,728 cubic millimeters. What is the volume of a similar cone with dimensions that are one-eighth the dimensions of the original? (Example 2) _____

6. A triangular prism has a volume of 350 cubic meters. If the dimensions are tripled, what is the volume of the new prism? (Example 2) _____

7. The model of a new apartment building is shown. The architect plans for the building to be 144 times the dimensions of the model. What will be the volume and surface area of the new building when it is completed? (Example 3)

_____

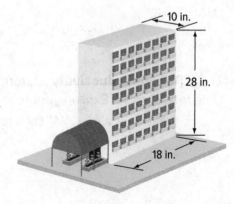

10 in.

28 in.

18 in.

**8.** The world's largest cube puzzle is in Knoxville, Tennessee. It measures 6 feet on each side. The scale factor between a standard cube puzzle and the largest puzzle is $\frac{1}{24}$. Find the surface area and volume of the standard cube puzzle. (Example 3)

_____

**9**  **Persevere with Problems** Two spheres are similar in shape. The scale factor between the smaller sphere and the larger sphere is $\frac{3}{4}$. If the volume of the smaller sphere is 126.9 cubic meters, what is the volume of the larger sphere? _____

---

## H.O.T. Problems  Higher Order Thinking

**10.**  **Persevere with Problems**  A *frustum* is the solid left after a cone is cut by a plane parallel to its base and the top cone is removed.

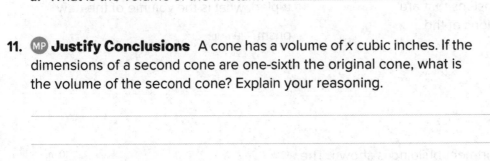

  **a.** Is the smaller cone that is removed similar to the original cone? Justify your response. _____

  **b.** What is the volume of the smaller cone? the larger cone? Use 3.14 for π. _____

  **c.** What is the ratio of the volume of the smaller cone to the volume of the larger cone? _____

  **d.** What is the volume of the frustum? _____

**11.**  **Justify Conclusions**  A cone has a volume of $x$ cubic inches. If the dimensions of a second cone are one-sixth the original cone, what is the volume of the second cone? Explain your reasoning.

_____

_____

_____

**12.**  **Reason Inductively**  Determine whether the following statement is *true* or *false*. Explain your reasoning.

  *All spheres are similar.*

_____

_____

# Extra Practice

**Copy and Solve  For Exercises 13–29, show your work and answers on a
separate piece of paper.**

**13** The surface area of a triangular prism is 300 square feet. What is the
surface area of a similar prism with dimensions that are 3 times greater than
the dimensions of the original prism?

**14.** The surface area of a rectangular prism is 1,350 square inches. What is
the surface area of a similar prism with dimensions that are 2 times greater
than the dimensions of  the original prism?

**15.** A pyramid has a volume of 640 cubic centimeters. If the dimensions of the
pyramid are reduced to one-fourth of the original dimensions, what is the
volume of the new pyramid?

**16.** The surface area of a rectangular prism is 1,300 square inches. Find
the surface area of a similar solid that is larger by a scale factor of 3.

**17.** The surface area of a triangular prism is 10.4 square meters. What is the
surface area of a similar solid that is  smaller by a scale factor of $\frac{1}{4}$?

**18.** Find the missing measures for the pair of
similar solids.

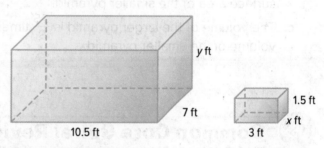

**MP** **Reason Inductively**  Determine whether each statement is *always*,
*sometimes*, or *never* true.

**19.** Two prisms with congruent bases are similar.

**20.** Similar solids have equal volumes.

**21.** Two cubes are similar.

**22.** A prism and pyramid are similar.

**23.** Two similar cylinders are shown.

  **a.** What is the ratio of their radii?

  **b.** What is the ratio of their surface areas? volumes?

  **c.** Find the surface area of Cylinder B.

  **d.** Find the volume of Cylinder A.

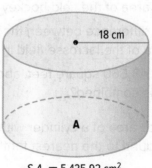

**24.** The side lengths of Cube *A* are 3 times the side length of Cube *B*. The side lengths of Cube *B* are half the side lengths of Cube *C*. Select the correct values to complete the following statements.

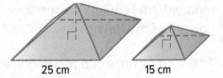

| $\frac{1}{16}$ | $\frac{1}{8}$ | $\frac{1}{4}$ | $\frac{1}{2}$ |
|---|---|---|---|
| 2 | 3 | 4 | 6 |
| 8 | 9 | 16 | 27 |

   **a.** The surface area of Cube *A* is ☐ times the surface area of Cube *B*.

   **b.** The volume of Cube *A* is ☐ times the volume of Cube *B*.

   **c.** The volume of Cube *B* is ☐ the volume of Cube *C*.

   **d.** The surface area of Cube *B* is ☐ the surface area of Cube *C*.

**25.** The two pyramids shown are similar.

Determine if each statement is true or false.

25 cm      15 cm

   **a.** The scale factor from the smaller pyramid to the larger pyramid is $\frac{5}{3}$.     ☐ True   ☐ False

   **b.** The surface area of the larger pyramid is $\frac{5}{3}$ times the surface area of the smaller pyramid.     ☐ True   ☐ False

   **c.** The volume of the larger pyramid is $\frac{25}{9}$ times the volume of the smaller pyramid.     ☐ True   ☐ False

---

 **Common Core Spiral Review**

**26.** Julianna is making a clay figurine of a dog. The dog is 75 centimeters tall. If she uses a scale of 1 centimeter = 10 centimeters, how tall will the clay figurine be? **7.G.1**

**27.** The table shows the dimensions of the fields used in various sports. **7.G.6**

   **a.** What is the area of the field hockey field in square feet?

   **b.** What is the difference between the area of the soccer field and the area of the lacrosse field in square feet?

   **c.** If an acre is 43,560 square feet, about how many acres are all four fields combined?

| Sport | Length (yards) | Width (yards) |
|---|---|---|
| Field hockey | 60 | 100 |
| Football | $53\frac{1}{3}$ | 120 |
| Lacrosse | 60 | 110 |
| Soccer | 70 | 115 |

**28.** Find the surface area of a cylinder with a radius of 15 meters and a height of 5 meters. Round to the nearest tenth. **8.G.9**

**29.** Find the surface area of a cone with a diameter of 4.5 centimeters and a slant height of 12 centimeters. Round to the nearest tenth. **8.G.9**

# 21ST CENTURY CAREER
## in Architecture

## Space Architect

Do you like building things? Are you an excellent problem solver? If so, you have what it takes to be a space architect. Space architects use principles from architecture, design, engineering, and science to create places for people to live and work in outer space. Their designs include transfer vehicles, lunar habitats, and Martian greenhouses. Because of the limitations, space architecture must be very efficient and functional. Every square inch of surface and every cubic inch of space must have a purpose.

**College & Career** READINESS

Explore college and careers at ccr.mcgraw-hill.com

## Is This the Career for You?

Are you interested in a career as a space architect? Take some of the following courses in high school.

◆ Aerospace Technology
◆ Calculus
◆ Geometry
◆ Introductory Space Planning
◆ Intro to CAD

Turn the page to find out how math relates to a career in Architecture.

## MP Out of this World Architecture

**Use the space laboratories below to solve each problem. Round to the nearest tenth.**

1. *Destiny* has one round window that is 20 inches in diameter. What is the circumference and area of the window?

   _____

2. What is the volume of *Destiny*? _____

3. The internal volume of *Columbus*, or the space where the astronauts live and work, is about 34.7 cubic meters less than the total volume. What is the internal volume

   of *Columbus*? _____

4. Find the surface area of *Destiny*. _____

5. Without calculating, predict whether *Destiny* or *Columbus* has a greater surface area. Then test your prediction by calculating

   the solution. _____

   _____

   _____

6. *Kibo* is a Japanese laboratory on the International Space Station. It is a cylinder 11.2 meters long with a radius of 2.2 meters. Compare its volume to the volumes of *Destiny*

   and *Columbus*. _____

   _____

   _____

**DESTINY**

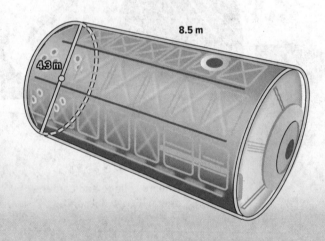

8.5 m

4.3 m

**COLUMBUS**

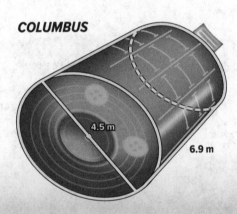

4.5 m

6.9 m

## MP Career Project

It's time to update your career portfolio! Use the Internet or another source to research a career as a space architect. Write a paragraph that summarizes your findings.

_____

_____

_____

_____

_____

*What subject in school is the most important to you? How would you use that subject in this career?*

_____

_____

_____

_____

# Chapter Review

## Vocabulary Check

**Complete each sentence using the vocabulary list at the beginning of the chapter.**

1. A set of all points in space that are a given distance from a given

   point is called a _____.

2. Cubes are _____ because they have the same
   shape and their corresponding linear measures are proportional.

3. A _____ is a three-dimensional figure with
   two parallel congruent circular bases connected by a curved surface.

**Reconstruct the vocabulary word and definition from the letters under the grid. The letters for each column are scrambled directly under that column.**

4.

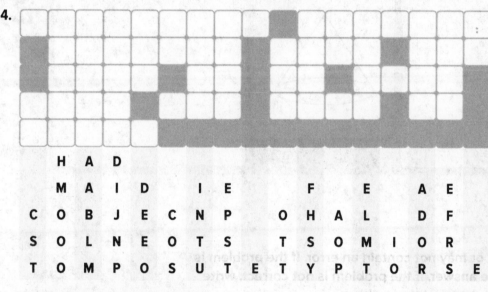

```
      H  A  D
      M  A  I  D      I  E        F     E     A  E
   C  O  B  J  E  C  N  P     O  H  A  L     D  F
   S  O  L  N  E  O  T  S     T  S  O  M  I  O  R
   T  O  M  P  O  S  U  T  E  T  Y  P  T  O  R  S  E
```

5.

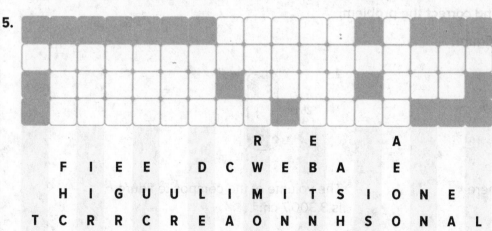

```
                        R     E        A
      F  I  E  E     D  C  W  E  B  A     E
      H  I  G  U  U  L  I  M  I  T  S  I  O  N  E
   T  C  R  R  C  R  E  A  O  N  N  H  S  O  N  A  L
```

## Use Your FOLDABLES

**Use your Foldable to help review the chapter.**

Tape here ↓                                                                     Tape here ↓

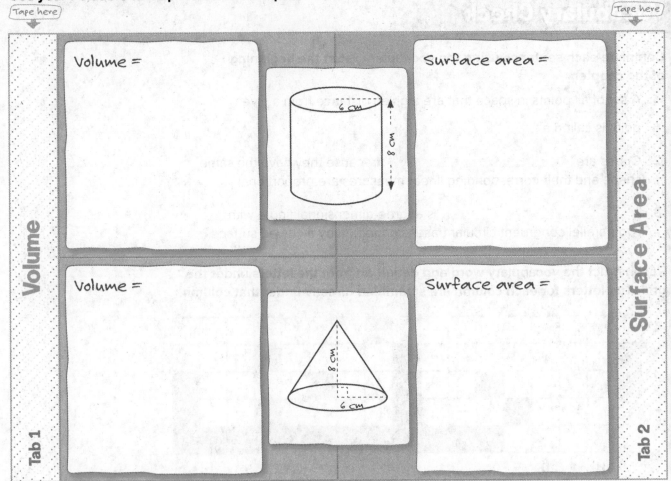

**Volume**

Tab 1

Volume =

Volume =

Surface area =

Surface area =

**Surface Area**

Tab 2

## Got it?

The problems below may or may not contain an error. If the problem is correct, write a "✓" by the answer. If the problem is not correct, write an "X" over the answer and correct the problem.

**Find the volume of each figure.**

1.

14 mm

The volume of the sphere is 11,494.0 mm³.

2.

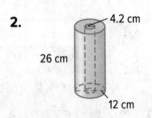

4.2 cm

26 cm

12 cm

The volume of the composite figure is 3,300.7 cm³.

# CCSS Power Up! Performance Task

## Popcorn Packaging

Midvale Theater is considering using cylinders or cones to serve popcorn. The measurements of the proposed cylinder are shown.

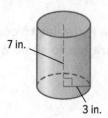

7 in.

3 in.

**Write your answers on another piece of paper. Show all of your work to receive full credit. Use 3.14 for $\pi$.**

### Part A

The manager of the theater wants to sell the popcorn in a cone with a diameter of 7 inches. What does the height of the cone need to be in order to hold the same amount of popcorn as the cylinder? Round your answer to the nearest tenth.

### Part B

The monthly popcorn special comes in a bowl that is shaped like a hemisphere. The diameter of the bowl is 10 inches. The theatre sells the cylindrical container of popcorn for $5.25. If they sell the bowl of popcorn for $5.00, is it a good deal? Explain your reasoning.

### Part C

Suppose the popcorn containers cost $0.002 per square inch to manufacture. Is it less expensive to produce the cone in Part A or the cylinder? Assume that neither container comes with a lid. Explain your reasoning.

**Answering the Essential Question**

Use what you learned about volume and surface area formulas to complete the graphic organizer. Name four topics that use a formula to solve a problem.

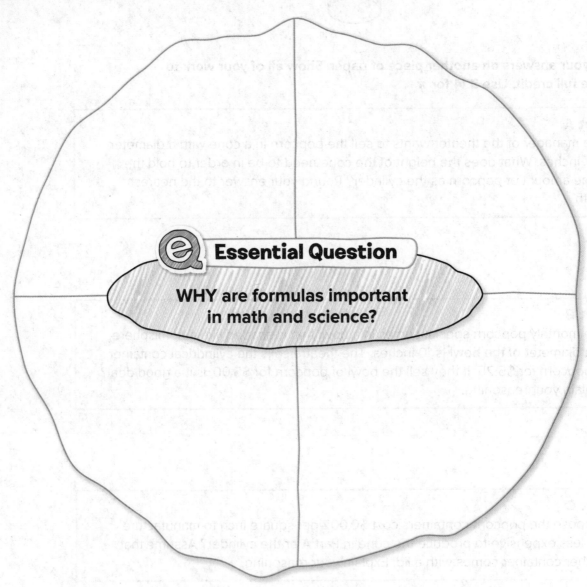

**Essential Question**

WHY are formulas important in math and science?

 **Answer the Essential Question.** WHY are formulas important in math and science?

_____

_____

# UNIT PROJECT

**Watch ▶**

**Design That Ride** Designers apply many geometric concepts to build exciting new rides. In this project you will:

- **Collaborate** with your classmates as you research amusement park rides.
- **Share** the results of your research in a creative way.
- **ℯ Reflect** on how you use different measurements to solve real-life problems.

By the end of this project, you'll be ready to design a new amusement park ride!

## Collaborate

**Collaborate**

**⏻ Go Online** Work with your group to research and complete each activity. You will use your results in the Share section on the following page.

1. Find photos of several different types of amusement park rides. Then describe and label any parallel lines, angles, triangles, congruent figures, similar figures, and three-dimensional shapes you observe.

2. Make a sketch of a current amusement park ride. Research and label its dimensions. Then use what you know about angles, congruence, similarity, the Pythagorean Theorem, surface area, and volume to label as many of its other attributes as you can.

3. Find several examples of amusement park rides that use transformations. Explain the transformation(s) that the ride exhibits.

4. Research different types of trusses. Then explain why trusses are used in the design of some amusement park rides. Include drawings to justify your explanation.

5. Research *potential energy* and *kinetic energy* as they relate to roller coasters. Then create a drawing that explains the concepts.

With your group, decide on a way to share what you have learned about designing an amusement park ride. Some suggestions are listed below, but you can also think of other creative ways to present your information. Remember to show how you used mathematics to complete each of the activities in this project.

- Design an amusement park ride using online simulations. Don't forget to give your ride a name.
- Imagine a nearby amusement park is seeking suggestions for a new ride. Write a proposal for your ride. Be sure to include drawings.

Check out the note on the right to connect this project with other subjects.

**connect** with **Social Studies**

**Global Literacy** Research information about amusement parks in other countries. Some questions to consider are:

- How are the rides different from those in the United States?
- When and where was the earliest amusement park constructed?

6. **Ⓠ Answer the Essential Question** HOW can you use different measurements to solve real-life problems?

   a. How did you use what you learned about the measurements of triangles to solve real-life problems in this project?

   _____

   _____

   b. How did you use what you learned about the measurements involved in congruence and similarity to solve real-life problems in this project?

   _____

   _____

   c. How did you use what you learned about volume and surface area to solve real-life problems in this project?

   _____

   _____

# UNIT 5

## Statistics and Probability

## Essential Question

**WHY is learning mathematics important?**

### Chapter 9
### Scatter Plots and Data Analysis

Scatter plots can be used to investigate patterns of association between two quantities. In this chapter, you will construct and interpret scatter plots. You will also draw lines of best fit to model data that suggest a linear association.

# Unit Project Preview

Watch

**Olympic Games** Basketball has been a Summer Olympic sport since 1936. You can compare and analyze certain data and statistics based on the scores from various games. The scores from other sports in the Summer Olympics can also be compared and analyzed in the same way.

At the end of Chapter 9, you'll complete a project to learn about scoring in basketball and other Olympic sports. But for now, it's time to do an activity in your book. Name some Summer Olympic sports you enjoy watching and explain their scoring systems.

 Summer Olympic Sports

# Chapter 9
# Scatter Plots and Data Analysis

## Essential Question

HOW are patterns used when comparing two quantities?

## Common Core State Standards

**Content Standards**
8.SP.1, 8.SP.2, 8.SP.3, 8.SP.4

**MP Mathematical Practices**
1, 2, 3, 4, 5, 7

## Math in the Real World

**Weather** During any given year, the Earth is struck by lightning more than 100 times every second. The numbers of lightning strikes for some states in a recent time period were 10, 10, 11, 12, 12, 14, 18, 20, 20, 21, 39, and 47.

Complete the histogram using the data.

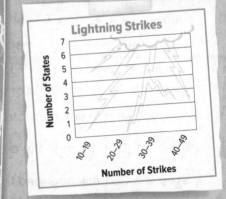

## FOLDABLES®
### Study Organizer

**1** Cut out the Foldable on page FL11 of this book.

**2** Place your Foldable on page 728.

**3** Use the Foldable throughout this chapter to help you learn about scatter plots and two-way tables.

# What Tools Do You Need?

 **Vocabulary**

| | |
|---|---|
| bivariate data | relative frequency |
| distribution | scatter plot |
| five-number summary | standard deviation |
| line of best fit | symmetric |
| mean absolute deviation | two-way table |
| qualitative data | univariate data |
| quantitative data | |

## Study Skill: Reading Math

**Topic Sentences** A topic sentence is a sentence that expresses the main idea in a paragraph. In a word problem, the "topic sentence" is usually found near the end. It is the sentence or question that tells you what you need to find. The "topic sentence" in the following example is underlined for you.

> Mrs. Garcia's math class was doing research about wild horses living on public lands. They found that there are about 30,000 wild horses living in Nevada, 4,000 living in Wyoming, and 2,000 living in California. <u>Is the number of wild horses living on public lands in Nevada, Wyoming, and California greater than 35,000?</u>

When you start to solve a word problem, follow these steps.

**Step 1** Skim through the problem, looking for the "topic sentence."

**Step 2** Go back and read the problem more carefully, looking for the supporting details you need to solve the problem.

For each exercise, <u>underline</u> the "topic sentence". Do not solve the problem.

1. Mirielle collected data for her science fair project on the relationship between a person's arm span and their height. She wanted to determine whether a relationship exists.

2. The two-way table shows the places that males and females volunteered in the past month. Do a higher percentage of males or females volunteer at the animal shelter?

# What Do You Already Know?

Place a checkmark below the face that expresses how much you know about each concept. Then scan the chapter to find a definition or example of it.

😞 I have no clue.　　　😐 I've heard of it.　　　😊 I know it!

| Concept | 😞 | 😐 | 😊 | Definition or Example |
|---|---|---|---|---|
| bivariate data | | | | |
| five-number summaries | | | | |
| lines of best fit | | | | |
| positive and negative associations | | | | |
| scatter plots | | | | |
| two-way tables | | | | |

Integers

# When Will You Use This?

Here are a few examples of how unit rates are used in the real world.

**Activity 1** How much is the cost of a movie ticket at your local movie theater? Has the price gone up recently? Does it cost more to see a 3-D movie? If so, how much more?

_____

_____

**Activity 2** Go online at **connectED.mcgraw-hill.com** to read the graphic novel **Movie Mania**. What do Danielle and Enrique want to find out about movie prices?

_____

Danielle and Enrique in
Movie Mania
SUMMER 2014

Try the Quick Check below.
Or, take the Online Readiness Quiz.

**Common Core Review** 6.SP.4, 6.SP.5

## Example 1

The ages of people in a play are shown in the histogram. Describe the histogram. Then find the number of play performers under the age of 30.

There are 4 + 5 + 5 + 1 + 1 or 16 performers. Most of the performers are between the ages of 20 and 39.

There are 4 + 5 or 9 performers under the age of 30.

## Example 2

On five plays, the Hawks completed passes for 15, 3, 8, 4, and 5 yards. What was the average number of yards per completed pass?

Find the sum of the numbers. Then divide by how many numbers are in the set.

$$\frac{15 + 3 + 8 + 4 + 5}{5} = \frac{35}{5}$$
$$= 7$$

The Hawks gained an average of 7 yards per completed pass.

### Quick Check

1. **Graphs** Describe the histogram. Then find the number of mammals with a life span of more than 20 years.

_____

_____

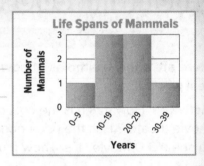

**Data Analysis** Find the mean (average) for each data set. Round to the nearest tenth if necessary.

2. 14, 17, 20, 16, 13 _____

3. 52, 36, 17, 41, 18, 29, 28, 32 _____

4. In 12 games last season the school baseball team scored 5, 11, 2, 0, 4, 8, 9, 6, 7, 4, 1, and 2 runs. What is the average number of runs scored per game? Round to the nearest tenth. _____

### How Did You Do?

Which problems did you answer correctly in the Quick Check? Shade those exercise numbers below.

① ② ③ ④

# Inquiry Lab
## Scatter Plots

 **Inquiry** HOW can I use a graph to investigate the relationship or trends between two sets of data?

 **Content Standards** 8.SP.1

**MP Mathematical Practices** 1, 3, 5

Mirielle collected data for her science fair project on the relationship between a person's arm span and their height. She wanted to determine whether or not a relationship exists.

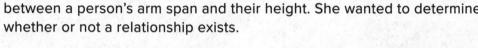 **Hands-On Activity**

Sometimes it is difficult to determine whether a relationship exists between two sets of data simply by looking at them. You can write the data as a set of ordered pairs and graph them on a coordinate plane.

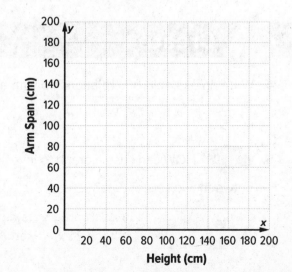

 **Step 1** Have a classmate measure your height and the length of your arm span with a meter stick to the nearest centimeter. Then write your height *x* and arm span *y* as an ordered pair. ( ___ , ___ )

**Step 2** Combine your data with that of your classmates. Write the ordered pairs (height, arm span) in the table.

| Classmates' Data (height, arm span) | | | | | |
|---|---|---|---|---|---|
| | | | | | |
| | | | | | |
| | | | | | |
| | | | | | |

**Step 3** Graph the ordered pairs (height, arm span) on the coordinate plane.

Is there a noticeable trend in the data? If so, describe the trend as *positive* or *negative*.

_____

_____

Using your graph, estimate the arm spans of two people whose heights are 150 centimeters and 185 centimeters.

_____

## Investigate

Collaborate

**Work with a partner.**

1. To determine whether a relationship exists between the circumference and the diameter of a circle, find 6 different circular objects in the room.

   a. Measure and record the diameter and circumference of each object in inches.

   b. Write the measurements of each circle as an ordered pair (d, C). Graph the ordered pairs on the coordinate plane.

| | diameter, d | circumference, C | (d, C) |
|---|---|---|---|
| 1 | | | |
| 2 | | | |
| 3 | | | |
| 4 | | | |
| 5 | | | |
| 6 | | | |

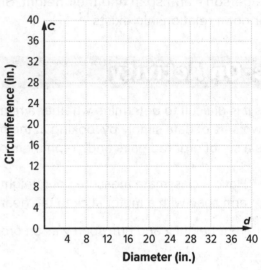

## Analyze and Reflect

Collaborate

2. **MP Reason Inductively** Is there a noticeable trend in the data? If so, describe the trend as *positive* or *negative*. Explain. _____

   _____

3. **MP Use Math Tools** Use your graph to estimate the circumference C of a circle with a diameter d of 10 inches. _____

## Create

On Your Own

4. Write an example of a relationship with a negative association.

   _____

   _____

5. **inquiry** HOW can I use a graph to investigate the relationship or trends between two sets of data? _____

   _____

   _____

# Scatter Plots

## Vocabulary Start-Up

Recall that the graph of a linear equation is a line on the coordinate plane. The slope of the line describes the direction and steepness of the line.

**On the coordinate grid shown, graph and label two lines. One line should have a positive slope and one line should have a negative slope.**

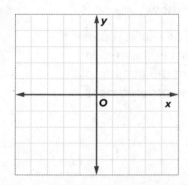

 ## Real-World Link

 Watch ▶

**Weather** The table shows temperatures in degrees Celsius and the corresponding temperatures in Fahrenheit from a local weather station. Graph the ordered pairs (°C, °F). Is the slope of the line passing through the points positive

or negative? _____

| °C | 0 | 5 | 10 | 15 | 20 | 25 | 30 |
|----|----|----|----|----|----|----|----|
| °F | 32 | 41 | 50 | 59 | 68 | 77 | 86 |

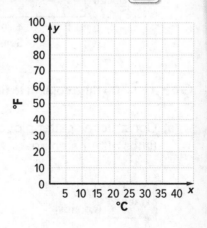

 **Essential Question**

HOW are patterns used when comparing two quantities?

 **Vocabulary**

bivariate data
scatter plot

**CCSS** **Common Core State Standards**

**Content Standards**
8.SP.1
**MP** **Mathematical Practices**
1, 3, 4

**Which MP Mathematical Practices did you use?**
**Shade the circle(s) that applies.**

① Persevere with Problems
② Reason Abstractly
③ Construct an Argument
④ Model with Mathematics

⑤ Use Math Tools
⑥ Attend to Precision
⑦ Make Use of Structure
⑧ Use Repeated Reasoning

# Construct a Scatter Plot

Data with two variables, or pairs of numerical observations, are called **bivariate data**. A **scatter plot** shows the relationship between bivariate data graphed as ordered pairs on a coordinate plane. For example, the bivariate data set year and number of visitors can be displayed as a scatter plot.

## Example

1. **Construct a scatter plot of the number of viewers who watched new seasons of a certain television show.**

   Let the horizontal axis, or *x*-axis, represent the number of seasons. Let the vertical axis, or *y*-axis, represent the number of viewers. Then graph the ordered pairs (season, viewers).

| Television Ratings | |
|---|---|
| Season | Viewers (millions) |
| 1 | 31.7 |
| 2 | 26.3 |
| 3 | 25.0 |
| 4 | 24.7 |
| 5 | 22.6 |
| 6 | 22.1 |

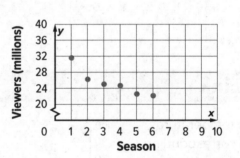

**Got it?** Do this problem to find out.

a. Construct a scatter plot of the weight of an alligator at various times after hatching.

| Weeks | Weight (pounds) |
|---|---|
| 0 | 6 |
| 9 | 8.6 |
| 18 | 10 |
| 27 | 13.6 |
| 34 | 15 |
| 43 | 17.2 |
| 49 | 19.8 |

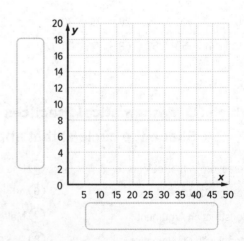

# Types of Associations

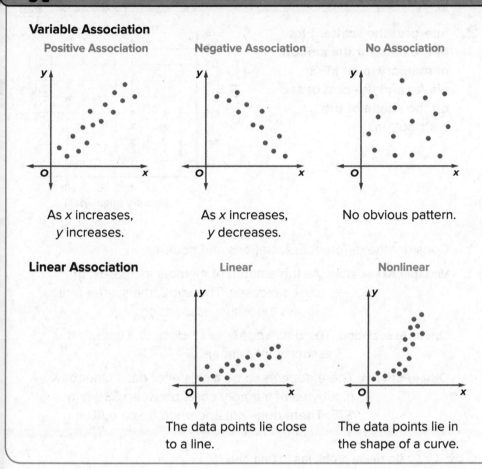

**Variable Association**

| Positive Association | Negative Association | No Association |
|---|---|---|
| As *x* increases, *y* increases. | As *x* increases, *y* decreases. | No obvious pattern. |

**Linear Association**

| Linear | Nonlinear |
|---|---|
| The data points lie close to a line. | The data points lie in the shape of a curve. |

You can analyze the shape of the distribution of a scatter plot to investigate patterns of association. If the distribution shows a positive or negative association, then the distribution can be classified as linear or nonlinear. The scatter plot below shows a positive nonlinear association. Clusters or outliers can also be identified.

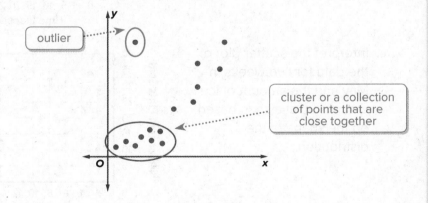

outlier

cluster or a collection of points that are close together

Tutor

## Example

**2.** Interpret the scatter plot of the data for the amount of memory in an MP3 player and the cost based on the shape of the distribution.

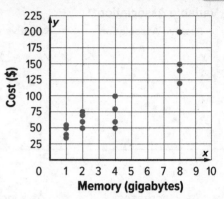

Consider the different associations and patterns.

**Variable Association**  As the amount of memory increases, the cost increases. Therefore, the scatter plot shows a positive association.

**Linear Association**  The data appear to lie close to a line, so the association is linear.

**Other Patterns**  There appears to be a cluster of data. One to two gigabytes of memory costs between $30 and $75. There does not appear to be an outlier.

Show your work.

**Got it?**  Do these problems to find out.

**b.** Interpret the scatter plot of the data for the time elapsed and temperature of water based on the shape of the distribution.

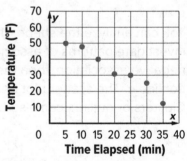

b. _____

**c.** Interpret the scatter plot of the data for two weeks in May and the amount of ice cream sold at a shop based on the shape of the distribution.

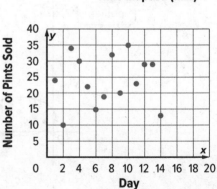

c. _____

## Example

Tutor

**3.** The table shows public school enrollment from 1999–2010.

| Years Since 1999 | 0 | 1 | 2 | 3 | 4 | 5 |
|---|---|---|---|---|---|---|
| Number of Students (millions) | 46.9 | 47.2 | 47.7 | 48.2 | 48.5 | 48.8 |

| Years Since 1999 | 6 | 7 | 8 | 9 | 10 | 11 |
|---|---|---|---|---|---|---|
| Number of Students (millions) | 49.1 | 49.3 | 49.3 | 49.6 | 49.8 | 50.0 |

**Construct and interpret a scatter plot of the data. If an association exists, make a conjecture about the number of students that will be enrolled in public schools in the year 2015.**

Construct a scatter plot of the data. Let the horizontal axis represent the year since 1999 and the vertical axis represent the number of students.

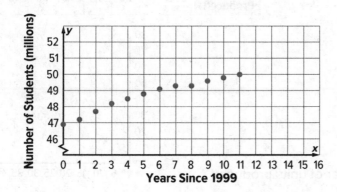

Consider the different associations and patterns.

**Variable Association** As the years increase, the number of students increases. Therefore, the scatter plot shows a positive association.

**Linear Association** The data appear to lie close to a line, so the association is linear.

**Other Patterns** There are no clusters or outliers.

To make a conjecture about the number of students that will be enrolled in public schools in the year 2015, follow the pattern until the x-value is 16. Then find the corresponding y-value.

So, there will be about 51 million students enrolled in public schools in 2015.

**Got it?** Do this problem to find out.

d. Interpret the scatter plot shown for the men's Olympic 100-meter freestyle swim winning times. If an association exists, make a conjecture about the winning time in the 2016 Olympics.

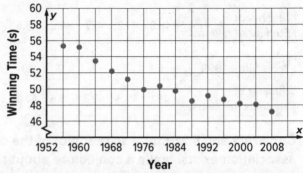

# Guided Practice

1. The table shows the number of units produced in a certain number of hours at a manufacturing plant. (Examples 1–3)

| Time (h) | 8 | 19 | 16 | 40 | 34 | 8 | 40 | 19 | 34 |
|---|---|---|---|---|---|---|---|---|---|
| Units Produced | 20 | 41 | 28 | 60 | 49 | 28 | 63 | 40 | 58 |

a. Construct a scatter plot of the data.

b. Interpret the scatter plot of the data.

_____

_____

_____

c. Make a conjecture about the number of units produced

in 50 hours. _____

2.  **Building on the Essential Question** What are the inferences that can be drawn from sets of data points having a positive association and a negative association?

_____

_____

_____

_____

_____

_____

_____

## Rate Yourself!

How confident are you about creating and interpreting scatter plots? Check the box that applies.

For more help, go online to access a Personal Tutor.

 Tutor

**FOLDABLES** Time to update your Foldable!

# Independent Practice

Go online for Step-by-Step Solutions

**1** Construct a scatter plot of the number of books donated over time. (Example 1)

| Year | 1 | 2 | 3 | 4 | 5 | 6 | 7 | 8 |
|---|---|---|---|---|---|---|---|---|
| **Number of Books** | 27 | 38 | 24 | 47 | 58 | 65 | 63 | 68 |

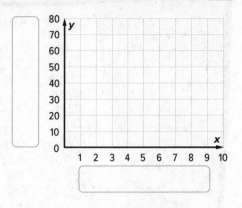

2. Interpret the scatter plot of the data for the amount of paint used to paint signs of various lengths based on the shape of the distribution. (Example 2) _____

_____

_____

_____

_____

_____

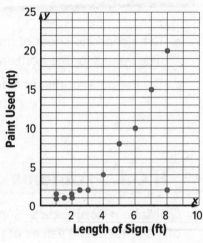

3. The table shows the amount of time different students studied for a test and their test scores. (Example 3)

| Time (min) | 10 | 15 | 20 | 25 | 30 | 35 | 40 | 45 |
|---|---|---|---|---|---|---|---|---|
| **Test Score** | 65 | 68 | 67 | 78 | 79 | 85 | 89 | 92 |

a. Construct a scatter plot of the data.

b. Interpret the scatter plot of the data based on the shape of the distribution.

_____

_____

_____

c. If a relationship exists, make a conjecture about the test score for a student who studied for

60 minutes. _____

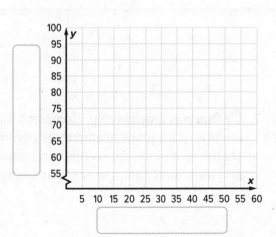

4. **MP Model with Mathematics** Refer to the graphic novel frame below for Exercises a–b.

a. On a separate sheet of grid paper, construct a scatter plot of the data. The values for the horizontal axis should be years since 1995.

b. Do the data represent a *positive, negative,* or *no* association? Explain.

_____

## H.O.T. Problems Higher Order Thinking

5. **MP Make a Conjecture** Suppose a scatter plot shows that as the values of *x* decrease, the values of *y* decrease. Does the scatter plot show

a *positive, negative,* or *no* association? _____

6. **MP Persevere with Problems** Is it *always, sometimes,* or *never* true that a scatter plot that shows a positive association suggests that the relationship is

proportional? Justify your answer. _____

_____

_____

7. **MP Reason Inductively** Complete the table that shows the side lengths of a square related to its perimeter and area. Would a scatter plot of the side length and perimeter or the side length and area represent a linear relationship? Explain.

| Side Length (units) | Perimeter (units) | Area (units²) |
|---|---|---|
| 1 | | |
| 2 | | |
| 3 | | |
| 4 | | |
| 5 | | |
| 6 | | |

_____

_____

# Extra Practice

**Copy and Solve  For Exercises 8–16, show your work and answers on a separate piece of paper.**

8. Construct and interpret a scatter plot of the data collected by a travel agency. If a relationship exists, make a conjecture about the number of visitors in month 12.

| Month | 1 | 2 | 3 | 4 | 5 | 6 | 7 | 8 | 9 | 10 |
|---|---|---|---|---|---|---|---|---|---|---|
| Number of Visitors | 208 | 245 | 423 | 432 | 412 | 626 | 647 | 620 | 402 | 356 |

9. The table shows the number of junk E-mails Petra received over the last 10 days.

| Day | 1 | 2 | 3 | 4 | 5 | 6 | 7 | 8 | 9 | 10 |
|---|---|---|---|---|---|---|---|---|---|---|
| Number of E-Mails | 10 | 12 | 15 | 10 | 11 | 8 | 20 | 10 | 10 | 9 |

  a. Construct a scatter plot of the data.

  b. Interpret the scatter plot of the data based on the shape of the distribution.

  c. If a relationship exists, make a conjecture about the number of junk E-mails on Day 15.

**Explain whether the scatter plot of the data for each of the following shows a *positive*, *negative*, or *no* association.**

10.
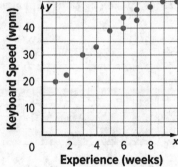

11.
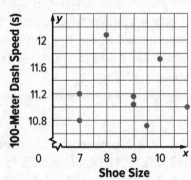

12. **MP Multiple Representations**  A thirteen-year-old takes an average of 14 breaths per minute.

  a. **Tables**  Let *x* represent minutes and *y* represent the number of times the teenager breathes. Make a table using the *x*-values of 1, 2, 3, 4, 8, and 10.

  b. **Graphs**  Make a scatter plot of the data. Describe the association between minutes and the number of times a person breathes.

  c. **Words**  Predict how many times a person would breathe in 25 minutes. Explain your reasoning.

13. The scatter plot shows the relationship between the average number of hours spent typing per week and the number of words typed per minute. Circle the appropriate word in the statement below to draw an accurate conclusion about the relationship shown in the scatter plot.

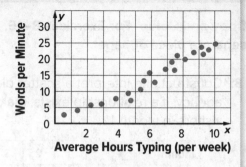

As the average number of hours spent typing per week (increases, decreases) the number of words typed per minute (increases, decreases).

14. The table shows the cost of fruit sold at a produce stand. Construct a scatter plot of the data.

| Fruit (lb) | 2 | 4 | 6 | 8 | 10 |
|---|---|---|---|---|---|
| Cost ($) | 5.00 | 10.00 | 12.00 | 15.00 | 16.00 |

Write a statement that is supported by the scatter plot.

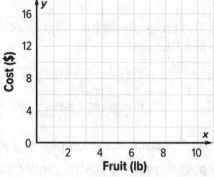

15. The graph shows the top five languages spoken by at least 100 million native speakers worldwide. What conclusions can you make about the number of Mandarin native speakers and the number of English native speakers? 6.SP.4

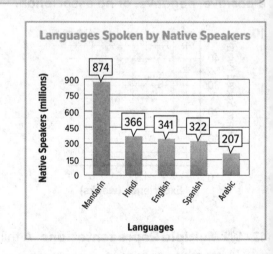

16. For a school food drive, Homeroom 212 collected 8, 17, 4, 10, 8, 8, 12, 20, 10, 11, 12, 13, and 25 food items. Homeroom 215 collected 16, 24, 10, 15, 12, 14, 12, 30, 15, 10, 15, 20, and 14 food items. 6.SP.4

a. Construct a double box plot for the data.

b. Compare the donations of the two homerooms.

# Inquiry Lab
## Lines of Best Fit

 **Inquiry** HOW can I use a data model to predict an outcome?

 **Content Standards**
8.SP.2

**MP Mathematical Practices**
1, 3

Sashi and Stacie found the data below showing the winning times in the Olympics for the women's 100-meter freestyle swim. They want to predict the winning time in the 2024 Olympics.

| 100-Meter Freestyle Winning Times | | | | | | | |
|---|---|---|---|---|---|---|---|
| Years Since 1956 | 0 | 4 | 8 | 12 | 16 | 20 | 24 |
| Winning Time (s) | 62.0 | 61.2 | 59.5 | 60.0 | 58.59 | 55.65 | 54.79 |
| Years Since 1956 | 28 | 32 | 36 | 40 | 44 | 48 | 52 |
| Winning Time (s) | 55.92 | 54.93 | 54.65 | 54.5 | 53.83 | 53.84 | 53.12 |

# Hands-On Activity

**Step 1** Construct a scatter plot by graphing the points (years since 1956, time).

**Step 2** Use a piece of uncooked spaghetti to make a line that goes through most of the data points.

How close are the other data points to the line you drew?

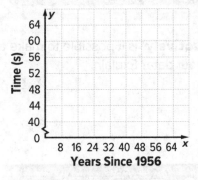

**Step 3** Look at the point where the spaghetti has an *x*-value of 68. The corresponding *y*-value represents the projected winning time in 2024.

What is the projected winning time in 2024? _____

Refer to the line drawn in the scatter plot. Is this method always valid when making a prediction?

_____

_____

**Investigate**

Collaborate

**Work with a partner.**

1. Research and collect a set of data from a newspaper or the Internet that has a positive or negative association.

   **a.** Compile your data in the table below. Space has been provided for ten sets of data values. Use a separate sheet of paper if you need more room. Be sure to label the rows.

   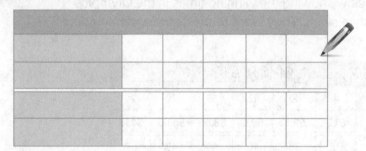

   **b.** Construct a scatter plot for the data by graphing your data as ordered pairs.

   **c.** Draw a line that goes through most of the data points.

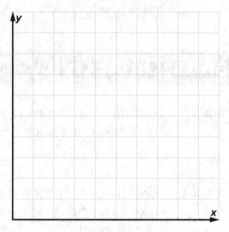

**Analyze and Reflect**

Collaborate

2. **MP Reason Inductively** Is it possible to make a prediction from your data? Explain.

_____

_____

**Create**

On Your Own

3. **MP Model with Mathematics** Create a set of data for which a prediction cannot be made.

_____

_____

4. **Inquiry** HOW can I use a data model to predict an outcome?

_____

_____

_____

# Lines of Best Fit

## Real-World Link

**Cookies** The table shows the average annual cost of one pound of chocolate chip cookies.

| Years Since 2000 | 0 | 1 | 2 | 3 | 4 | 5 | 6 | 7 | 8 | 9 |
|---|---|---|---|---|---|---|---|---|---|---|
| Average Cost ($) | 2.59 | 2.81 | 2.65 | 2.67 | 2.88 | 2.70 | 2.85 | 2.88 | 3.17 | 3.24 |

1. What year corresponds to 0 years since 2000? _____

   9 years since 2000? _____

2. If the data were displayed in a scatter plot, would the scatter plot show a positive, negative, or no association? Explain.

   _____

   _____

3. Would a more reasonable prediction for the cost of cookies in 2015 be $3.25 or $4.00? Explain.

   _____

   _____

   _____

### Essential Question

HOW are patterns used when comparing two quantities?

**Vocabulary**

line of best fit

### Common Core State Standards

**Content Standards**
8.SP.1, 8.SP.2, 8.SP.3

**Mathematical Practices**
1, 3, 4, 5

Which **MP** **Mathematical Practices** did you use?
Shade the circle(s) that applies.

① Persevere with Problems  ⑤ Use Math Tools

② Reason Abstractly  ⑥ Attend to Precision

③ Construct an Argument  ⑦ Make Use of Structure

④ Model with Mathematics  ⑧ Use Repeated Reasoning

# Line of Best Fit

When data are collected, the points graphed usually do not form a straight line, but may approximate a linear relationship. A **line of best fit** is a line that is very close to most of the data points.

 **Examples**

Tutor

Refer to the information in the table about the cost of cookies.

**STOP and Reflect**

How do you determine how well a line of best fit models the data? Explain below.

**1.** Construct a scatter plot using the data. Then draw and assess a line that seems to best represent the data.

| Years Since 2000 | 0 | 1 | 2 | 3 | 4 | 5 | 6 | 7 | 8 | 9 |
|---|---|---|---|---|---|---|---|---|---|---|
| Average Cost ($) | 2.59 | 2.81 | 2.65 | 2.67 | 2.88 | 2.70 | 2.85 | 2.88 | 3.17 | 3.24 |

Graph each of the data points. Draw a line that fits the data. About half of the points are above the line and half of the points are below the line. Judge the closeness of the data points to the line. Most of the points are close to the line.

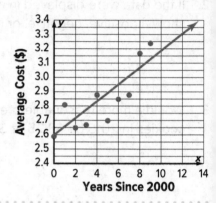

**2.** Use the line of best fit to make a conjecture about the cost of cookies in 2013.

Extend the line so that you can estimate the $y$-value for an $x$-value of 2013 − 2000 or 13. The $y$-value for 13 is about $3.35. We can predict that in 2013, a pound of chocolate chip cookies will cost $3.35.

**Got it?** Do these problems to find out.

Refer to the scatter plot about yearly points scored by a certain race car driver.

**a.** Draw and assess a line that seems to best represent the data.

**b.** Use the line of best fit to make a conjecture about the points the driver will score in 2015.

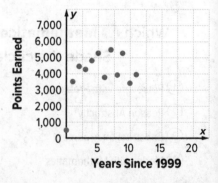

a. _____

b. _____

Show your work.

# Examples

Tutor

The scatter plot shows the number of cellular service subscribers in the U.S.

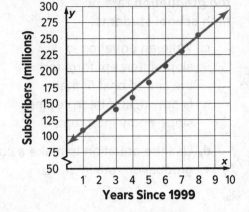

**Subscribers (millions)**

**Years Since 1999**

**3.** Write an equation in slope-intercept form for the line of best fit that is drawn, and interpret the slope and y-intercept.

Choose any two points on the line. They may or may not be data points. The line passes through points (3, 150) and (9, 275). Use these points to find the slope, or rate of change, of the line.

$m = \dfrac{y_2 - y_1}{x_2 - x_1}$     Definition of slope

$m = \dfrac{275 - 150}{9 - 3}$     $(x_1, y_1) = (3, 150)$ and $(x_2, y_2) = (9, 275)$

$m = \dfrac{125}{6}$ or about 20.83     Simplify.

The slope is about 20.83. This means the number of cell phone subscribers increased by about 20.83 million people per year.

The y-intercept is 87.5 because the line of fit crosses the y-axis at about the point (0, 87.5). This means there were about 87.5 million cell phone subscribers in 1999.

$y = mx + b$     Slope-intercept form

$y = 20.83x + 87.5$     Replace m with 20.83 and b with 87.5.

The equation for the line of best fit is $y = 20.83x + 87.5$.

**Estimation**

Drawing a line of best fit using the method in this lesson is an estimation. Therefore, it is possible to draw different lines to approximate the same data.

**4.** Use the equation to make a conjecture about the number of cellular subscribers in 2015.

The year 2015 is 16 years after 1999.

$y = 20.83x + 87.5$     Equation for the line of best fit

$y = 20.83(16) + 87.5$     Replace x with 16.

$y = 333.28 + 87.5$     Simplify.

So, in 2015, there will be about 420.83 million cellular subscribers.

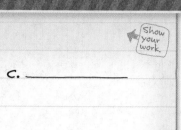

c. _____

d. _____

**Got it?** Do these problems to find out.

The scatter plot shows the graduation rate of high school students.

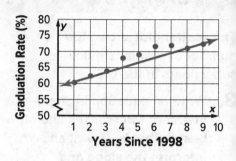

c. Write an equation in slope-intercept form for the line of best fit that is drawn, and interpret the slope and y-intercept.

d. Use the equation to make a conjecture about the graduation rate in 2020.

## Guided Practice

Check ✓

1. The table shows the life expectancy, in years, for people born in certain years.
(Examples 1–4)

| Years Since 1900 | 0 | 10 | 20 | 30 | 40 | 50 | 60 | 70 | 80 | 90 | 100 |
|---|---|---|---|---|---|---|---|---|---|---|---|
| Life Expectancy | 47.3 | 50.0 | 54.1 | 59.7 | 62.9 | 68.2 | 69.7 | 70.8 | 73.7 | 75.4 | 77.1 |

a. Construct a scatter plot of the data. Then draw and assess a line that best represents the data.

_____

_____

_____

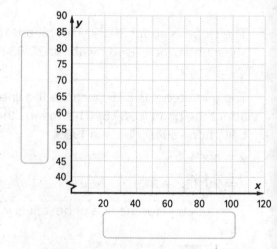

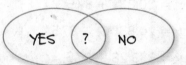

b. Write an equation in slope-intercept form for the line of fit and interpret the slope and y-intercept.

_____

_____

c. Use the equation to make a conjecture about the life expectancy for a person born in 2020.

_____

2. ⓔ **Building on the Essential Question** Why do we estimate a line of best fit for a scatter plot?

_____

_____

_____

**Rate Yourself!**

Are you ready to move on?
Shade the section that applies.

YES   ?   NO

For more help, go online to access a Personal Tutor.

Tutor

**FOLDABLES** Time to update your Foldable!

# Independent Practice

Go online for Step-by-Step Solutions

eHelp

**1** The results of a survey about women's shoe sizes and heights are shown. (Examples 1 and 2)

a. Construct a scatter plot of the data. Then draw and assess a line that best represents the data.

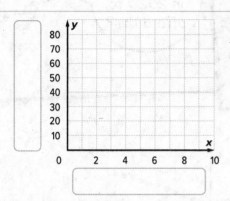

Show your work.

| Height (inches) and Shoe Size | | | |
|---|---|---|---|
| Shoe Size | Height | Shoe Size | Height |
| 8 | 66 | $6\frac{1}{2}$ | 65 |
| 8 | 65 | 9 | 68 |
| $7\frac{1}{2}$ | 65 | $7\frac{1}{2}$ | 63 |
| 7 | 62 | 7 | 64 |
| 7 | 62 | $5\frac{1}{2}$ | 62 |
| 9 | 68 | 5 | 60 |
| 9 | 65 | 9 | 67 |
| 9 | 65 | 6 | 59 |

b. Use the line of best fit to make a conjecture about the height of a female who wears a size 5 shoe. _____

2. The table shows the number of Calories burned when walking laps around a track. (Examples 1-4)

| Laps Completed | 1 | 2 | 3 | 4 | 5 | 6 | 7 |
|---|---|---|---|---|---|---|---|
| Calories Burned | 30 | 70 | 80 | 112 | 150 | 170 | 225 |

a. Construct a scatter plot of the data. Then draw a line that best represents the data.

b. Write an equation for the line of best fit. Use the equation to make a conjecture about the number of Calories burned if someone walks 15 laps.

_____

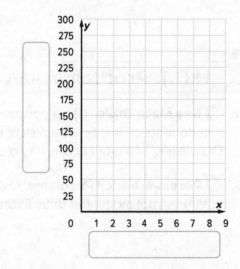

**3** The scatter plot shows the number of girls who participate in ice hockey. (Examples 3 and 4)

a. Write an equation in slope-intercept form for the line of best fit that is drawn, and interpret the slope and *y*-intercept. _____

_____

_____

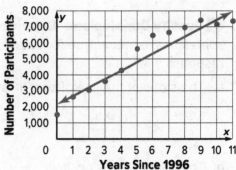

b. Use the equation to make a conjecture about the number of girls that will participate in ice hockey in 2020. _____

4. **MP** **Model with Mathematics** Refer to the graphic novel frame below for Exercises a–b.

a. The scatter plot shows the average ticket prices since 1995. Draw a line that best represents the data in your scatter plot.

b. Write an equation in slope-intercept form for the line of best fit. Make a conjecture about the cost of a movie ticket in 2020.

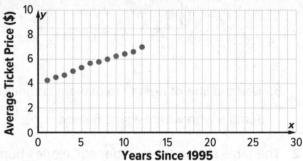

## 🔥 H.O.T. Problems Higher Order Thinking

5. **MP** **Use Math Tools** Use a newspaper or the Internet to find a scatter plot that consists of at least seven data points. Tape the scatter plot to your book. Draw a line of best fit and write an equation for the line.

6. **MP** **Persevere with Problems** Describe or draw a scatter plot where a line of fit does not model the data. Explain your reasoning to a classmate.

7. **MP** **Justify Conclusions** Determine whether each statement is *always*, *sometimes*, or *never* true for data with a positive association. Justify your response.

a. The slope of the line of best fit is positive.

b. The *y*-intercept is positive.

# Extra Practice

**Copy and Solve** **For Exercises 8–14, show your work and answers on a separate piece of paper.**

8. The table shows the storage amount and cost for flash drives at a local electronics store.

| Storage (MB) | 2 | 2 | 2 | 2 | 4 | 4 | 4 | 4 | 8 | 8 | 8 | 8 | 8 | 16 | 16 | 16 | 16 |
|---|---|---|---|---|---|---|---|---|---|---|---|---|---|---|---|---|---|
| Cost ($) | 6.5 | 12 | 7 | 10 | 9 | 10 | 20 | 23 | 15 | 17 | 20 | 25 | 40 | 23 | 40 | 50 | 80 |

a. Construct a scatter plot of the data. Draw and assess a line that best represents the data.

b. Write an equation in slope-intercept form for the line of best fit, and interpret the slope and *y*-intercept.

c. Use the equation to make a conjecture about the cost of a flash drive with a storage capacity of 32 MB.

9. **MP** **Model with Mathematics** The table shows fat and Calories for fast food sandwiches.

| Fat (grams) | 21 | 10 | 14 | 21 | 30 | 34 | 32 | 37 | 27 | 26 | 18 | 7 |
|---|---|---|---|---|---|---|---|---|---|---|---|---|
| Calories | 490 | 280 | 330 | 430 | 530 | 590 | 540 | 590 | 550 | 470 | 450 | 340 |

a. Construct a scatter plot of the data. Draw and assess a line that best represents the data.

b. Write an equation in slope-intercept form for the line of best fit, and interpret the slope and *y*-intercept.

c. Use the equation to make a conjecture about the number of grams of fat in a sandwich with 350 Calories.

10. The table shows the cost per pound of apples for several years.

| Years Since 1999 | 1 | 2 | 3 | 4 | 5 | 6 | 7 | 8 | 9 | 10 | 11 |
|---|---|---|---|---|---|---|---|---|---|---|---|
| Cost per Pound ($) | 0.92 | 0.87 | 0.95 | 0.98 | 1.04 | 1.07 | 1.12 | 1.12 | 1.32 | 1.18 | 1.22 |

a. Construct a scatter plot of the data. Draw and assess a line that best represents the data.

b. Write an equation in slope-intercept form for the line of best fit, and interpret the slope and *y*-intercept.

c. Use the equation to make a conjecture about the cost of apples in the year 2025.

**11.** Kayla recorded data on how many Calories she burned for different lengths of time while jogging on a treadmill. She plotted the data in a scatter plot and drew the line of best fit. The equation for the line is $C = 14.5m$, where $C$ represents the number of Calories burned and $m$ represents the number of minutes spent jogging. Determine if each statement is true or false.

    **a.** The slope is positive because as Kayla jogs more minutes, she burns more Calories.    ☐ True    ☐ False

    **b.** According to the line of best fit, Kayla will burn about 290 Calories if she jogs 20 minutes.    ☐ True    ☐ False

**12.** The table shows the wind chill temperatures for different wind speeds when the outside temperature is 30°F. Construct a scatter plot of the data. Then draw a line of best fit.

| Wind Chill Temperatures at 30°F | | | |
|---|---|---|---|
| Wind Speed (mph) | Temperature (°F) | Wind Speed (mph) | Temperature |
| 5 | 25 | 25 | 16 |
| 10 | 21 | 30 | 15 |
| 15 | 19 | 35 | 14 |
| 20 | 17 | 40 | 13 |

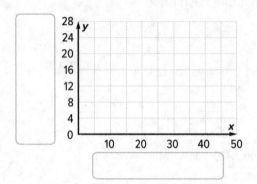

Does the line of best fit have a positive or negative slope? Explain what this represents.

**Determine whether a scatter plot of each of the following might show a positive, negative, or no association.** 8.SP.1

**13.** a student's age and how many siblings he or she has

**14.** the number of homeruns hit and the amount of time spent in batting practice

# Inquiry Lab

## Graphing Technology: Linear and Nonlinear Association

 **Inquiry**

**HOW can you use technology to describe associations in scatter plots?**

 **Content Standards**
8.SP.1, 8.SP.2, 8.SP.3

 **Mathematical Practices**
1, 3, 5

The table shows the weekly number of hours spent watching television and the weekly number of hours spent exercising.

| Weekly Television (h) | 17 | 20 | 11 | 10 | 15 | 38 | 5 | 25 |
|---|---|---|---|---|---|---|---|---|
| Weekly Exercise (h) | 5 | 4.5 | 7.5 | 8 | 6.5 | 1 | 7.5 | 3 |
| Weekly Television (h) | 25 | 32 | 5 | 17 | 40 | 28 | 20 | 30 |
| Weekly Exercise (h) | 2.5 | 3.5 | 6 | 7 | 0.5 | 5 | 4 | 1.5 |

# Hands-On Activity 1

You can use a graphing calculator to construct a scatter plot of the data and find and graph a line of best fit.

**Step 1**   Clear the existing data by pressing ⎣STAT⎦ ⎣ENTER⎦ ▲ ⎣CLEAR⎦ ⎣ENTER⎦. Then enter the data. Input the number of weekly hours spent watching television in $L_1$ and press ⎣ENTER⎦. Then enter the weekly hours spent exercising in $L_2$.

**Step 2**   Turn on the statistical plot by pressing ⎣2nd⎦ ⎣STAT PLOT⎦ ⎣ENTER⎦ ⎣ENTER⎦. Select the scatter plot and confirm $L_1$ as the Xlist, $L_2$ as the Ylist, and the square as the mark.

**Step 3**   Graph the data by pressing ⎣ZOOM⎦ 9. Use the Trace feature and the left and right arrow keys to move from one point to another.

Do the data suggest a linear association? _____

**Step 4**   Access the CALC menu by pressing ⎣STAT⎦ ▶. Select 4 to find a line of best fit in the form $y = ax + b$. Press ⎣2nd⎦ ⎣L1⎦ , ⎣2nd⎦ ⎣L2⎦ ⎣ENTER⎦ to find a line of best fit for the data in lists $L_1$ and $L_2$. What does the screen show for y, a, and b?

$y = $ _____   $a = $ _____   $b = $ _____

**Step 5** Graph the line of best fit in $Y_1$ by pressing $\boxed{Y=}$ and then $\boxed{\text{VARS}}$ 5 to access the Statistics... menu. Use the $\boxed{\blacktriangleright}$ and $\boxed{\text{ENTER}}$ keys to select EQ. Then press 1 to select RegEQ, the line of best fit equation. Finally, press $\boxed{\text{GRAPH}}$.

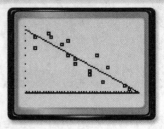

Use the TRACE feature to predict the average number of hours of exercise someone who watches 35 hours of television would get.

# Hands-On Activity 2

A nonlinear association is one where the pattern does not follow a linear trend.

The table shows the side length and the corresponding area for various squares. Construct a scatter plot of the data to determine what kind of relationship, if any, exists between the side of a square and its area.

| Side Length (cm) | Area (cm$^2$) |
|---|---|
| 0.5 | 0.25 |
| 1 | 1 |
| 1.5 | 2.25 |
| 2 | 4 |
| 2.5 | 6.25 |
| 3 | 9 |
| 3.5 | 12.25 |

**Step 1** Clear the equation from $Y_1$ by pressing $\boxed{Y=}$ $\boxed{\text{CLEAR}}$. Clear the existing data from $L_1$ and $L_2$ by pressing $\boxed{\text{STAT}}$ $\boxed{\text{ENTER}}$ $\boxed{\blacktriangle}$ $\boxed{\text{CLEAR}}$ $\boxed{\text{ENTER}}$ $\boxed{\blacktriangleright}$ $\boxed{\blacktriangle}$ $\boxed{\text{CLEAR}}$ $\boxed{\text{ENTER}}$.

**Step 2** Next, enter the data. Input the side lengths in $L_1$ and press $\boxed{\text{ENTER}}$. Then enter the areas in $L_2$.

**Step 3** Turn on the statistical plot by pressing $\boxed{\text{2nd}}$ $\boxed{\substack{\text{STAT} \\ \text{PLOT}}}$ $\boxed{\text{ENTER}}$ $\boxed{\text{ENTER}}$. Select the scatter plot and confirm $L_1$ as the Xlist, $L_2$ as the Ylist, and the square as the mark.

**Step 4** Graph the data by pressing $\boxed{\text{ZOOM}}$ 9. Use the Trace feature and the left and right arrow keys to move from one point to another.

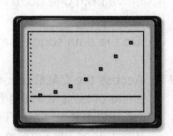

Does the scatter plot show a *linear* or *nonlinear* association? Explain.

## Investigate

Collaborate

**Work with a partner.**

Camryn lives in Missouri. She kept track of how much her energy bill was every month for one year. She displayed it in the table shown at the right. Use your graphing calculator and the following steps to construct a scatter plot of the data.

| Month | Bill ($) |
|---|---|
| January | 146 |
| February | 138 |
| March | 116 |
| April | 84 |
| May | 72 |
| June | 73 |
| July | 94 |
| August | 114 |
| September | 92 |
| October | 91 |
| November | 126 |
| December | 139 |

**Step 1** Clear the existing data from $L_1$ and $L_2$.

**Step 2** Enter the data. Input the month numbers in $L_1$ and the amounts of the electric bill in $L_2$.

**Step 3** Turn on the statistical plot. Select the scatter plot and confirm $L_1$ as the Xlist, $L_2$ as the Ylist.

**Step 4** Graph the data.

1. How is the scatter plot different from the scatter plot for Activity 2?

2. Does the scatter plot show a *linear* or *nonlinear* association? Explain.

3. What does a negative rate of change mean in the problem's context?

4. **MP Use Math Tools** Collect a set of data that can be represented in a scatter plot. Use a graphing calculator to determine whether the data have a *linear* or *nonlinear* association. If the association is linear, use the graphing calculator to find the line of best fit and to make a prediction. Show your data and results in the space provided.

## Analyze and Reflect

**Work with a partner.**

The *correlation coefficient* measures the strength of the association between two sets of data, or how closely the data is clustered around the line of best fit.

You can use the graphing calculator to find the correlation coefficient for the data in Activity 1. Before you reenter the data, you need to make sure that you have the Diagnostics on. Press [2nd] [Catalog]. Scroll down until you see DiagnosticOn. Then press [ENTER] [ENTER].

Complete Steps 1–4 in Activity 1. This time, when you complete Step 4, you should see values for $r^2$ and r. The value for r is the correlation coefficient.

5. In Activity 1, what is the r value? _____

6. **MP** **Make a Conjecture** The chart shows how the r value reflects the strength of the association. For example, a strong negative association would indicate the data is tightly clustered around a line of fit with a negative slope.

| If... | $-1 \leq r \leq -0.5$ | $-0.5 < r < 0$ | $r = 0$ | $0 < r < 0.5$ | $0.5 \leq r \leq 1$ |
|---|---|---|---|---|---|
| ...then the association is... | strong negative | weak negative | no association | weak positive | strong positive |

How would you classify the association in Activity 1?

_____

## Create

**Write a correlation coefficient for each association. Explain why you selected each value.**

7. strong positive

_____

_____

_____

8. weak negative

_____

_____

_____

9. **Inquiry** HOW can you use technology to describe associations in scatter plots?

_____

_____

_____

# Two-Way Tables

**Student Athletes** The data from a survey of 440 students are shown in the table. The students were asked whether or not they were on the honor roll and whether or not they played a sport.

| Student Athlete Survey | |
| --- | --- |
| Only on the honor roll | 115 |
| Only play a sport | 45 |
| Play a sport and are on the honor roll | 250 |

1. Complete the Venn diagram to represent the data.

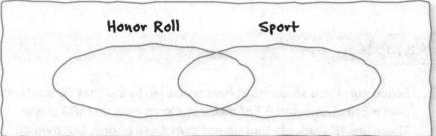

Honor Roll    Sport

2. **Relative frequency** is the ratio of the value of a subtotal to the value of the total. What is the relative frequency of a student that is on the honor roll and plays a sport to all students that are on the honor roll?

_____

3. Is there evidence that students that play sports are also on the honor roll? Explain.

_____

_____

_____

 **Essential Question**

HOW are patterns used when comparing two quantities?

 **Vocabulary**

relative frequency
two-way table

 **Common Core State Standards**

**Content Standards**
8.SP.4

**MP** **Mathematical Practices**
1, 3, 4, 5

Which **MP** **Mathematical Practices** did you use?
Shade the circle(s) that applies.

① Persevere with Problems    ⑤ Use Math Tools

② Reason Abstractly    ⑥ Attend to Precision

③ Construct an Argument    ⑦ Make Use of Structure

④ Model with Mathematics    ⑧ Use Repeated Reasoning

# Construct a Two-Way Table

A **two-way table** shows data from one sample group as it relates to two different categories.

The same information from the Venn diagram on the previous page is shown below as a two-way table, where one category is represented by rows and the other category is represented by columns. The two categories in the table shown are "Play a Sport" and "On the Honor Roll."

|  | Play a Sport | Do Not Play a Sport | Total |
|---|---|---|---|
| **On the Honor Roll** | 250 | 115 | 365 |
| **Not On the Honor Roll** | 45 | 30 | 75 |
| **Total** | 250 + 45 = 295 | 115 + 30 = 145 | 440 |

# Example

Tutor

1.  Felipe surveyed students at his school. He found that 78 students own a cell phone and 57 of those students own an MP3 player. There are 13 students that do not own a cell phone, but own an MP3 player. Nine students do not own either device. Construct a two-way table summarizing the data.

   **Step 1** Create a table using the two categories: cell phones and MP3 players. Fill in the table with the given values.

   |  | MP3 Player | No MP3 Player | Total |
   |---|---|---|---|
   | **Cell Phone** | 57 |  | 78 |
   | **No Cell Phone** | 13 | 9 |  |
   | **Total** |  |  |  |

   **Step 2** Use reasoning to complete the table. Remember, the totals are for each row and column. The column labeled "Total" should have the same sum as the row labeled "Total."

   |  | MP3 Player | No MP3 Player | Total |
   |---|---|---|---|
   | **Cell Phone** | 57 | 21 | 78 |
   | **No Cell Phone** | 13 | 9 | 22 |
   | **Total** | 70 | 30 | 100 |

## Got it? Do this problem to find out.

a. There are 150 children at summer camp and 71 signed up for swimming. There were a total of 62 children that signed up for canoeing and 28 of them also signed up for swimming. Construct a two-way table summarizing the data.

|  | Canoeing | No Canoeing | Total |
|---|---|---|---|
| Swimming |  |  |  |
| No Swimming |  |  |  |
| Total |  |  |  |

# Interpret Relative Frequencies

A two-way table can show relative frequencies for rows or for columns, rather than the actual values. By analyzing the relative frequencies in a two-way table, you can determine possible associations between the two variables.

## Example

Tutor

2. **Find and interpret the relative frequencies of students in the survey from Example 1 by row.**

|  | MP3 Player | No MP3 Player | Total |
|---|---|---|---|
| Cell Phone | 57 | 21 | 78 |
| No Cell Phone | 13 | 9 | 22 |
| Total | 70 | 30 | 100 |

To find the relative frequencies by row, write the ratios of each value to the total in that row. Round to the nearest hundredth.

|  | MP3 Player | No MP3 Player | Total |
|---|---|---|---|
| Cell Phone | $57; \frac{57}{78} \approx 0.73$ | $21; \frac{21}{78} \approx 0.27$ | 78; 1.00 |
| No Cell Phone | $13; \frac{13}{22} \approx 0.59$ | $9; \frac{9}{22} \approx 0.41$ | 22; 1.00 |

Only the totals needed are shown in the table.

Based on the relative frequency value of 0.73 in one of the cells, you can imply that most students that own a cell phone also own an MP3 player. The data also suggest that over half of the students that do not own a cell phone will own an MP3 player.

**STOP and Reflect**

What relative frequency would you use to determine if there was an association between the two variables in a two-way table? Explain below.

**b.** _____

## Got it? Do this problem to find out.

**b.** Find and interpret the relative frequencies of students in the survey by column. Round to the nearest hundredth if necessary.

|  | MP3 Player | No MP3 Player |
|---|---|---|
| **Cell Phone** | 57; | 21; |
| **No Cell Phone** | 13; | 9; |
| **Total** | 70; | 30; |

# Guided Practice

Check ✓

1. Eloise surveyed the students in her cafeteria and found that 38 males agree with the new cafeteria rules while 70 do not. There were 92 females surveyed and 41 of them agree with the new cafeteria rules. Construct a two-way table summarizing the data. (Example 1)

|  | Agree with Rules | Do Not Agree with Rules | Total |
|---|---|---|---|
| **Males** |  |  |  |
| **Females** |  |  |  |
| **Total** |  |  |  |

2. The two-way table shows how some students get their news. Find and interpret the relative frequencies of students in the survey by row. (Example 2)

|  | TV | Internet | Total |
|---|---|---|---|
| **7th grade** | 13; | 49; |  |
| **8th grade** | 20; | 68; |  |
| **Total** |  |  |  |

3.  **Building on the Essential Question** How is a two-way table used when determining possible associations between two different categories from the same sample group?

_____

_____

_____

_____

### Rate Yourself!

How well do you understand two-way tables? Circle the image that applies.

Clear    Somewhat Clear    Not So Clear

For more help, go online to access a Personal Tutor.

**FOLDABLES** Time to update your Foldable!

# Independent Practice

Go online for Step-by-Step Solutions    eHelp

**1** One hundred customers in a restaurant were asked whether they liked chicken or beef and whether they liked rice or pasta. Out of 30 customers that liked rice, 20 liked chicken. There were 60 customers that liked chicken. Construct a two-way table summarizing the data. (Example 1)

|  | Chicken | Beef | Total |
|---|---|---|---|
| **Rice** |  |  |  |
| **Pasta** |  |  |  |
| **Total** |  |  |  |

**2.** The two-way table shows the number of students that do or do not do chores at home and whether they receive an allowance or not. Find and interpret the relative frequencies of students in the survey by columns. (Example 2)

|  | Allowance | No Allowance | Total |
|---|---|---|---|
| **Chores** | 13; | 3; |  |
| **No Chores** | 5; | 4; |  |
| **Total** |  |  |  |

_____

_____

_____

**3** The two-way table shows the number of students that message on a daily basis. Find and interpret the relative frequencies of students in the survey by rows. (Example 2)

|  | Text Message | Instant Message | Total |
|---|---|---|---|
| **7th graders** | 59; | 25; |  |
| **8th graders** | 59; | 41; |  |
| **Total** |  |  |  |

_____

_____

**4.** **MP Use Math Tools** The Venn diagram shows the number of students that exercise in different ways. Construct a two-way table that displays the data. Find and interpret the relative frequencies by column.

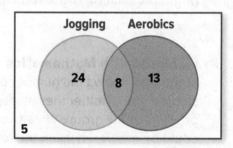

|  |  |  |
|---|---|---|
|  |  |  |
|  |  |  |
|  |  |  |

_____

_____

5. **MP Find the Error** Jasmine is interpreting data about her classmates that have an after-school job and are on the honor role. Out of 100 students that do not have a job, 67 of them are on the honor roll. Find her mistake and correct it.

*In my class, more than half of the students that are on the honor roll do not have after school jobs.*

_____

_____

_____

6. **MP Use Math Tools** Survey your classmates to find out if they have a curfew and if they have assigned chores at home. On a separate sheet of paper, make a two-way table that displays the results. Then interpret the results.

_____

7. **MP Persevere with Problems** The two-way table below shows the number of students with each hair color and eye color.

|  |  | Hair Color | | | | |
|---|---|---|---|---|---|---|
|  |  | Black | Brown | Red | Blond | Total |
| **Eye Color** | **Brown** | 7 | 12 | 3 | 1 | 23 |
|  | **Blue** | 2 | 8 | 2 | 9 | 21 |
|  | **Hazel** | 2 | 5 | 1 | 1 | 9 |
|  | **Green** | 1 | 3 | 1 | 2 | 7 |
|  | **Total** | 12 | 28 | 7 | 13 | 60 |

Which is greater: the percentage of the brown-haired students with blue eyes or the percentage of the red-haired students with brown eyes?

_____

8. **MP Model with Mathematics** The two-way table at the right shows the number of hours students studied and whether they studied independently or with a study group. Write two questions that could be answered using the relative frequencies of the data in the table. Then ask a classmate to solve your questions.

|  | Studied Less Than 2 Hours | Studied More Than 2 Hours |
|---|---|---|
| **Studied Independently** | 12 | 4 |
| **Studied with a Study Group** | 8 | 11 |

_____

_____

_____

_____

# Extra Practice

**Copy and Solve** For Exercises 9–17, show your work and answers on a separate piece of paper.

**9.** As each person entered the theater, Aaron counted how many of the 105 people had popcorn and how many had a drink. He found that out of 84 people that had popcorn, only 10 did not have a drink. Six people walked in without popcorn or a drink. Construct a two-way table summarizing the results.

**10.** The two-way table shows the number of Sasha's soccer teammates that are in her Math class and English class.

|  | Math Class | Not in Math Class |
|---|---|---|
| **English Class** | 4 | 2 |
| **Not in English Class** | 1 | 3 |

   **a.** How many teammates does Sasha have?

   **b.** What is the relative frequency of teammates that are in both of Sasha's classes to all of her teammates?

   **c.** Of the teammates in her math class, which percentage is greater: the percentage of teammates that are in her English class or the percentage of teammates that are not in her English class?

**11.** The two-way table shows the places that males and females volunteered in the past month. Do a greater percentage of males or females volunteer at the animal shelter? Justify your response.

|  | Males | Females |
|---|---|---|
| **Animal Shelter** | 26 | 21 |
| **Hospital** | 13 | 17 |
| **Library** | 9 | 14 |

**12.** **MP Use Math Tools** Cali surveyed the students in the cafeteria about the number of times they bring their lunch to school per month. The table shows her findings. Construct a two-way table that shows the relative frequencies by columns. What is the relative frequency of the number of girls that bring their lunch to school less than 6 times a month to the total number of students surveyed? Round to the nearest hundredth if necessary.

| Number of Times per Month | Males | Females |
|---|---|---|
| **0–5** | 35 | 25 |
| **6–10** | 23 | 16 |
| **11–15** | 22 | 13 |
| **16–20** | 18 | 8 |

**13.** A pet store conducted a survey on the types of pets owned by the store's customers. Construct a two-way table summarizing the data. Then find and interpret the relative frequencies by columns.

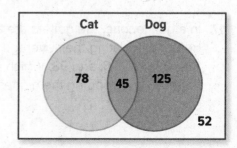

**14.** The Pep Club was asked to vote for which dinner they would like for the banquet. Complete the two-way table based on the information shown in the Venn diagram.

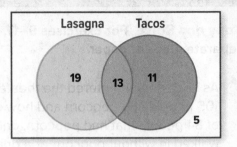

|  | Lasagna | No Lasagna | Total |
|---|---|---|---|
| **Tacos** |  |  |  |
| **No Tacos** |  |  |  |
| **Total** |  |  |  |

What is the relative frequency of students who chose tacos and lasagna to the total number of students? Round to the nearest hundredth.

[          ]

**15.** Megan surveyed the eighth grade students to find which school activities they attended last weekend. The results are shown in the two-way table. Which of the following are valid conclusions about the data? Select all that apply.

|  | Attended the School Play | Did Not Attend the School Play | Total |
|---|---|---|---|
| **Attended the Basketball Game** | 55 | 63 | 118 |
| **Did Not Attend the Basketball Game** | 88 | 15 | 103 |
| **Total** | 143 | 78 | 221 |

☐ Of the students who attended the basketball game, fewer than half of them also attended the school play.

☐ More than half of the students who were surveyed attended the school play and did not attend the basketball game.

☐ Students who attended the school play were more likely not to attend the basketball game.

## Common Core Spiral Review

**16.** The ages of people working in an office are shown in the stem-and-leaf plot. Find the mean, median, and mode of the data. 6.SP.5

**17.** In a golf league, the golfers are allowed to drop their highest score before calculating their average score. Seki has scores of 103, 98, 125, 96, 100, 95, and 98. Which measure of center will be affected the most by dropping the highest score? Explain. 6.SP.5

**Ages of Office Workers**

| Stem | Leaf |
|---|---|
| 2 | 3 5 8 8 |
| 3 | 1 2 3 3 6 9 |
| 4 | 2 5 7 |
| 5 | 1 3 |

2|3 = 23 years

# Problem-Solving Investigation
# Use a Graph

## Case #1 Up to Speed

 Watch

Marissa ranked ten Web sites from 1 to 10 with a ranking of 1 being the most popular. Then, she created a graph showing the download times of these Web sites.

*Does the most popular Web site have the fastest download time?*

**CCSS** Content Standards
8.SP.1

**MP** Mathematical Practices
1, 4, 7

**1** ## Understand *What are the facts?*

The graph shows the popularity of some Web sites and the download times for each.

**2** ## Plan *What is your strategy to solve this problem?*

Study the data on the graph.

**3** ## Solve *How can you apply the strategy?*

Use the graph to answer the following questions. The graph shows that, in general, the more popular

Web sites are _____ than the less popular Web sites.

The fastest Web site has what ranking? ☐

The slowest Web site has what ranking? ☐

**4** ## Check *Does the answer make sense?*

Look at the graph. Two Web sites have a higher rating than the fastest Web site.

## Analyze the Strategy  Tutor

**MP** **Identify Structure** Explain what the ordered pair (1, 1.4) represents in terms

of the question posed.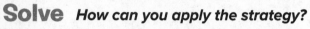

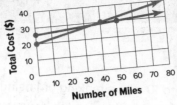

### Case #2 The Right Rental

The blue line shows the weekly cost of a car rental at Company A. The green line shows the weekly cost of a car rental at Company B.

*If you wish to rent a car for one week and drive 60 miles, which company charges the lesser amount?*

## Understand

**Read the problem. What are you being asked to find?**

I need to find _____

_____

**Underline key words and values. What information do you know?**

The graph shows the _____ and the _____ for

Companies A and B.

## Plan

**Choose a problem-solving strategy.**

I will use the _____ strategy.

## Solve

**Use your problem-solving strategy to solve the problem.**

The graph shows that at zero miles, Company A charges $☐ and

Company B charges $☐; but at 60 miles, Company A charges

about $☐ and Company B charges about $☐.

So, Company ☐ is less expensive to rent for one week and 60 miles.

## Check

**Use information from the problem to check your answer.**

Use the horizontal axis and find ☐ miles. Follow that vertical line up

to the car rental graphs.

The _____ line represents the less expensive car rental company.

**Work with a small group to solve the following cases.**
**Show your work on a separate piece of paper.**

## Case #3 Blogs

The numbers of followers of a popular blog are shown in the table.

What is a reasonable estimate for the number of followers in Year 10 if this trend continues?

_____

| Year | Number of Followers |
|------|---------------------|
| 1 | 42,000 |
| 2 | 50,000 |
| 3 | 76,000 |
| 4 | 94,000 |
| 5 | 115,000 |

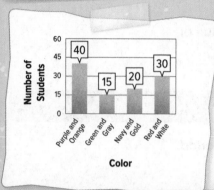

## Case #4 School Colors

The graph shows the results of a favorite color survey.

To the nearest percent, what percent more of the students chose purple and orange than green and gray?

_____

## Case #5 Clubs

The table shows the math club membership in the years 2010 through 2015.

What is a reasonable prediction for the membership in 2020 if this trend continues?

_____

| Year | Number of Members |
|------|-------------------|
| 2010 | 20 |
| 2011 | 21 |
| 2012 | 30 |
| 2013 | 34 |
| 2014 | 38 |
| 2015 | 45 |

## Case #6 Sales

A coat regularly sells for $125. The price is reduced by 10% each week for the next four weeks.

What is the mean price of the coat during this five-week period?

_____

Use any strategy!

# Mid-Chapter Check

## Vocabulary Check

1. **MP Be Precise** Define *bivariate data*. Give an example of a data set made up of bivariate data. (Lesson 1)

_____

_____

2. Fill in the blank with the correct term. (Lesson 3)

   The _____ in a two-way table is the ratio of the value of a subtotal to the value of the total.

## Skills Check and Problem Solving

The table below shows the average cost to own a certain car over a period of five years. (Lessons 1 and 2)

| Year | 1 | 2 | 3 | 4 | 5 |
|------|-----|-----|-----|-----|-----|
| Cost ($) | 10,600 | 7,900 | 8,000 | 8,100 | 7,000 |

3. Construct and interpret a scatter plot of the data.

   _____

   _____

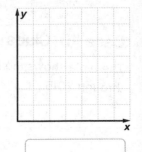

4. Draw a line of best fit.

5. Write an equation in slope-intercept form for the line of best fit and interpret the slope and *y*-intercept.

   _____

   _____

   _____

6. **MP Persevere with Problems** The two-way table shows the amount of time students studied for a test and the score they received. What is the relative frequency by column of the students that studied more than 30 minutes and received a score of 75% or more? (Lesson 3)

|  | Less than 30 minutes | More than 30 minutes |
|---|---|---|
| Score of 75% or more | 20 | 45 |
| Score below 75% | 33 | 27 |

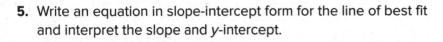

   _____

# Descriptive Statistics

## Vocabulary Start-Up

Recall that measures of center represent the middle of the data. The most common measures of center are mean, median, and mode.

**Complete the graphic organizer. Consider each word on the Knowledge Rating Scale and place a check ✓ in the appropriate column next to the word.**

| Knowledge Rating Scale | | | |
|---|---|---|---|
| Word | No clue | Have seen or heard it | Know it well |
| mean | | | |
| median | | | |
| mode | | | |

**Vocabulary**

univariate data
quantitative data
five-number summary

 **Common Core State Standards**

**Content Standards**
Preparation for S.ID.1 and S.ID.2

 **Mathematical Practices**
1, 2, 3, 4, 7

## Real-World Link

The data in the table represent the results of a survey about distance driven over spring break. Does the mean or the median best represent the data? Explain.

| Distance Driven over Spring Break (mi) | | | | |
|---|---|---|---|---|
| 749 | 312 | 302 | 296 | 293 |
| 277 | 257 | 256 | 219 | 209 |

_____

_____

_____

**Which MP Mathematical Practices did you use?**
**Shade the circle(s) that applies.**

① Persevere with Problems    ⑤ Use Math Tools

② Reason Abstractly    ⑥ Attend to Precision

③ Construct an Argument    ⑦ Make Use of Structure

④ Model with Mathematics    ⑧ Use Repeated Reasoning

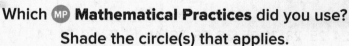

# Measures of Center and Variability

Data with one variable, such as test scores, are called **univariate data**. These data can be described by a measure of center.

## Example

**1.** The ages, in years, of the people seated in one row of a movie theater are 16, 15, 24, 33, 30, 56, 19, and 19. Find the mean, median, mode, and range of the data set.

**Mean** $\dfrac{16 + 15 + 24 + 33 + 30 + 56 + 19 + 19}{8} = \dfrac{212}{8}$ or 26.5

**Median** 15, 16, 19, $\underbrace{19,\ 24}$, 30, 33, 56    Arrange in order from least to greatest.

$\dfrac{19 + 24}{2} = 21.5$ years old

**Mode** The mode is 19, since it is the number that occurs most often.

**Range** $56 - 15 = 41$

### Got it? Do this problem to find out.

a. _____

**a.** Find the mean, median, mode, and range of the data set.

| Heights of Students (in.) | | | | |
|---|---|---|---|---|
| 66 | 72 | 70 | 74 | 64 |
| 65 | 60 | 62 | 66 | 67 |
| 68 | 71 | 70 | 72 | 73 |

# Five-Number Summary

**Quantitative data** are data that can be measured. A set of quantitative data can be divided into four equal parts, called quartiles.

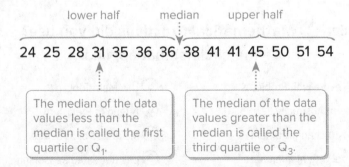

lower half        median        upper half

24  25  28  31  35  36  36  38  41  41  45  50  51  54

The median of the data values less than the median is called the first quartile or $Q_1$.

The median of the data values greater than the median is called the third quartile or $Q_3$.

This **five-number summary,** which includes the minimum value, first quartile ($Q_1$), median, third quartile ($Q_3$), and the maximum value of a data set, provides a numerical way of characterizing a set of data. The five-number summary can be described visually with a box plot, as shown below.

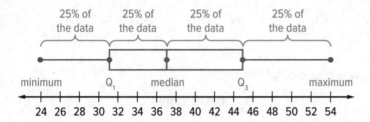

## Example

**2.** The data for daily average temperatures for 15 days in May are shown in the table.

| Temperature (°F) | | | | |
|---|---|---|---|---|
| 68 | 73 | 70 | 71 | 74 |
| 72 | 75 | 69 | 76 | 75 |
| 72 | 75 | 76 | 75 | 76 |

**a.** Find the five-number summary of the data.

Write the data from least to greatest.

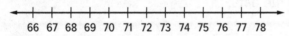

**68** 69 70 **71** 72 72 73 **74** 75 75 75 **75** 76 76 **76**

minimum    first quartile    median    third quartile    maximum

**b. Draw a box plot of the data.**

> **Step 1** Draw a number line that includes the least and greatest numbers in the data.

66 67 68 69 70 71 72 73 74 75 76 77 78

> **Step 2** Mark the minimum and maximum values, the median, and the first and third quartiles above the number line.

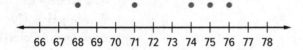

66 67 68 69 70 71 72 73 74 75 76 77 78

> **Step 3** Draw the box plot and assign a title to the graph.

**Daily Temperatures**

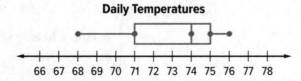

66 67 68 69 70 71 72 73 74 75 76 77 78

**Outliers**

If an asterisk (*) appears on a box plot, it represents an outlier. Outliers are data that are more than 1.5 times the interquartile range from the first or third quartiles.

b. _____

**Got it?** Do these problems to find out.

b. The points scored by a basketball team are shown in the table. Find the five-number summary of the data.

| Game | 1 | 2 | 3 | 4 | 5 | 6 | 7 | 8 | 9 |
|------|---|---|---|---|---|---|---|---|---|
| Number of Points | 34 | 20 | 83 | 36 | 37 | 44 | 40 | 35 | 36 |

c. Draw a box plot of the data.

**Points Scored**

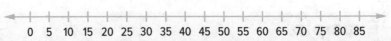

0  5  10  15  20  25  30  35  40  45  50  55  60  65  70  75  80  85

## Guided Practice  Check ✓

1. The points scored by each of seven basketball players is 12, 4, 18, 16, 21, 8, and 12. Find the mean, median, mode, and range of the data

 Show your work.

set. (Example 1) _____

_____

2. The data for Calories burned per minute of exercise is in the table. (Example 2)

| Exercise | Jogging | Jumping Rope | Basketball | Soccer | Bicycling | Downhill Skiing | Walking |
|----------|---------|--------------|------------|--------|-----------|-----------------|---------|
| Calories Burned | 8 | 7 | 7 | 6 | 5 | 5 | 4 |

a. Find the five-number summary of the data. _____

_____

b. Draw a box plot to represent the data.

**Calories Burned**

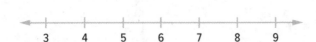

3  4  5  6  7  8  9

3.  e **Building on the Essential Question** What does the length of the "whiskers" in a box plot say about the data?

_____

_____

_____

**Rate Yourself!**

How confident are you about finding the five-number summary? Color the square that applies.

☐  ☐  ☐  ☐  ☐

For more help, go online to access a Personal Tutor.

Tutor 💬

# Independent Practice

Go online for Step-by-Step Solutions

**Find the mean, median, mode, and range of each data set. Round to the nearest tenth if necessary.** (Example 1)

**1.** Roller coaster speeds shown in the table at the right

| Fastest Roller Coasters | |
|---|---|
| **Coaster** | **Speed (mph)** |
| Dodonpa | 107 |
| Kingda Ka | 128 |
| Millennium Force | 93 |
| Phantom's Revenge | 82 |
| Steel Dragon 2000 | 95 |
| Superman: The Escape | 100 |
| Top Thrill Dragster | 120 |
| Tower of Terror | 100 |

**2.** Number of words in magazine articles: 115, 118, 115, 100, 97, 105

**Find the five-number summary of each set of data. Then draw a box plot of the data.** (Example 2)

**3.**

| Number of Days of Incubation Periods for Pet Birds | |
|---|---|
| Australian King Parrot | 20 |
| Glossy Cockatoo | 30 |
| Major Mitchell's Cockatoo | 26 |
| Princess Parrot | 21 |
| Red-Tailed Cockatoo | 30 |
| Red-Winged Parrot | 21 |
| Regent Parrot | 21 |
| Superb Parrot | 20 |
| White-Tailed Cockatoo | 29 |
| Yellow-Tailed Cockatoo | 29 |

**4.**

| Top Ten Countries Average Weekly Teen Spending | |
|---|---|
| Norway | $49.70 |
| Sweden | $41.70 |
| Brazil | $41.30 |
| Argentina | $40.50 |
| Hong Kong | $38.00 |
| United States | $37.60 |
| Denmark | $37.40 |
| Singapore | $34.10 |
| Greece | $32.90 |
| France | $31.30 |

**Incubation Period**

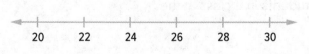

**Teen Spending**

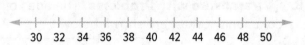

5. **MP Multiple Representations** A restaurant conducted a survey asking its customers to rate the new menu using a scale of 1 to 20. The results of the survey are shown in the line plot.

**Restaurant Survey Results**

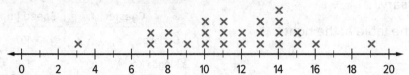

a. **Numbers** Find the mean, median, mode, and range of the data set. Round to the nearest tenth if necessary.

_____

_____

b. **Numbers** Find the five-number summary of the data set.

_____

_____

c. **Graphs** Draw a box plot to represent the set of data.

**Menu Survey**

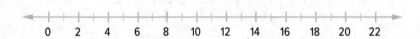

## H.O.T. Problems  Higher Order Thinking

6. **MP Reason Abstractly** Create a data set that contains 8 to 12 values such that the mean is greater than the median.

_____

7. **MP Persevere with Problems** Create two different data sets that have the same median and same quartiles, but different ranges.

_____

_____

8. **MP Persevere with Problems** The ages of the students in a class at the community center are shown below.

$$25, 28, 36, 21, 28, 15, 24, 30$$

If the age of the teacher is added to the set of data, the mean age becomes 27. What is the age of the teacher? _____

# Extra Practice

**Copy and Solve** For Exercises 9–16, show your work and answers on a separate piece of paper.

**Find the mean, median, mode, and range of each data set.**

**9** Length in inches of spools of ribbon: 60, 48, 36, 144, 72

**10.** Cost in dollars for medium pizza: 6, 6, 8, 10, 4, 6, 8, 9

**11.** Horace's bowling scores are shown in the table.

a. Find the mean, median, mode, and range of the data. Round to the nearest tenth.

| Bowling Scores | | | |
| --- | --- | --- | --- |
| 164 | 128 | 151 | 138 |
| 158 | 162 | 130 | 162 |

b. Find the five-number summary of the data.

c. Draw a box plot to represent the data.

**12.** The prices of digital cameras are shown in the table.

a. Find the mean, median, mode, and range of the data. Round to the nearest tenth.

| Prices of Digital Cameras ($) | | | | | |
| --- | --- | --- | --- | --- | --- |
| 100 | 180 | 250 | 200 | 130 | 180 |
| 250 | 280 | 90 | 300 | 300 | 750 |
| 150 | 130 | 200 | 180 | 100 | 350 |

b. Find the five-number summary of the data.

c. Draw a box plot to represent the data.

d. What conclusions can be drawn from the box plot?

**13.** **MP** **Identify Structure** Label the parts of the box plot.

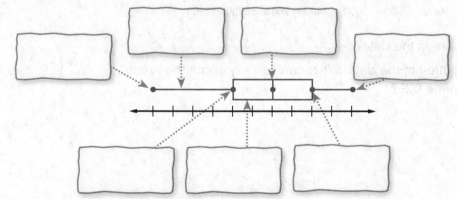

**14.** The table shows the ages of people standing in line for tickets to see a movie. Find each of the follow values for the data set.

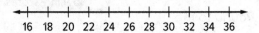

| Ages | | | | |
|---|---|---|---|---|
| 18 | 22 | 31 | 25 | 30 |
| 19 | 26 | 24 | 35 | 25 |

Minimum: ☐   First Quartile: ☐   Median: ☐

Third Quartile: ☐   Maximum: ☐

Draw a box plot to represent the set of data.

```
←┼──┼──┼──┼──┼──┼──┼──┼──┼──┼──┼──→
  16 18 20 22 24 26 28 30 32 34 36
```

**15.** The speeds, in miles per hour, of several cars on a busy street are shown. Determine if each statement is true or false.

| 42 | 38 | 44 | 35 | 50 | 38 |

  **a.** The range of the speeds is 12 miles per hour.     ☐ True   ☐ False

  **b.** The mean is the measure of center that makes the speeds appear fastest.     ☐ True   ☐ False

  **c.** The median is the measure of center that makes the speeds appear slowest.     ☐ True   ☐ False

## CCSS Common Core Spiral Review

**16.** The box plot below shows the areas of the largest zoos in the United States. 6.SP.5

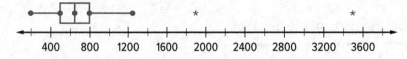

**Areas (acres) of the Ten Largest Zoos in the United States**

```
  ●──┬─┬──●        *              *
┼──┼──┼──┼──┼──┼──┼──┼──┼──┼──┼──→
400  800 1200 1600 2000 2400 2800 3200 3600
```

  **a.** How many outliers are in the data?

  **b.** Describe the distribution of the data. What can you say about the areas of the major zoos in the U.S.?

# Measures of Variation

## Real-World Link

**Restaurant** A restaurant asks the staff to record the number of people who order the special each day. The table shows the number of specials ordered per day.

| Day | 1 | 2 | 3 | 4 | 5 | 6 |
|---|---|---|---|---|---|---|
| Number of Specials | 26 | 25 | 30 | 32 | 27 | 28 |

**1.** Plot the data on the graph provided.

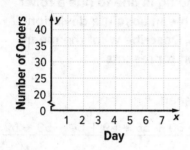

**2.** Find the mean of the data set. ☐

**3.** Complete the table by finding the absolute value of the difference between the mean and each data value in the set.

| Number of Specials | 26 | 25 | 30 | 32 | 27 | 28 |
|---|---|---|---|---|---|---|
| Difference from Mean | | | | | | |

**4.** Find the average of the values for the difference from the mean in the table. ☐

## Essential Question

HOW are patterns used when comparing two quantities?

**Vocabulary**

mean absolute deviation
standard deviation

**CCSS** **Common Core State Standards**

**Content Standards**
Preparation for S.ID.2
**MP** **Mathematical Practices**
1, 3, 4, 7

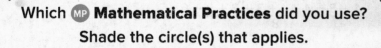

Which **MP** **Mathematical Practices** did you use?
Shade the circle(s) that applies.

① Persevere with Problems
② Reason Abstractly
③ Construct an Argument
④ Model with Mathematics
⑤ Use Math Tools
⑥ Attend to Precision
⑦ Make Use of Structure
⑧ Use Repeated Reasoning

*Work Zone*

# Mean Absolute Deviation

You have used measures of center to describe the middle of a set of data, and you have used range to describe the spread or *variation* of a set of data. Another way to describe the variation of a set of data is to use its mean absolute deviation. The **mean absolute deviation** of a set of data is the average distance between each data value and the mean.

## Example

Tutor

**1.** The table shows the heights of the first eight people standing in line to ride a roller coaster. Find the mean absolute deviation of the set of data. Describe what the mean absolute deviation represents.

| Heights (in.) | | | |
|---|---|---|---|
| 52 | 48 | 60 | 55 |
| 59 | 54 | 58 | 62 |

**Step 1**    Find the mean.

$$\frac{52 + 48 + 60 + 55 + 59 + 54 + 58 + 62}{8} = 56$$

**Step 2**    Find the absolute value of the differences between each value in the data set and the mean.

$|52 - 56| = 4$        $|59 - 56| = 3$

$|48 - 56| = 8$        $|54 - 56| = 2$

$|60 - 56| = 4$        $|58 - 56| = 2$

$|55 - 56| = 1$        $|62 - 56| = 6$

**Step 3**    Find the average of the absolute values of the differences between each value in the data set and the mean.

$$\frac{4 + 8 + 4 + 1 + 3 + 2 + 2 + 6}{8} = 3.75$$

The mean absolute deviation is 3.75. This means that the average distance each person's height is from the mean height is 3.75 inches.

## Got it? Do this problem to find out.

a. The number of points that Samantha scored in five basketball games was 8, 14, 10, 7, and 13. Find the mean absolute deviation of the set of data. Describe what the mean absolute deviation represents.

a. _____

# Standard Deviation

The **standard deviation** of a set of data is a calculated value that shows how the data deviates from the mean of the data. In a given set of data, most of the values fall within one standard deviation of the mean. So, if the mean of a set of data is 21 and the standard deviation is 3.5, then most of the values fall between 21 − 3.5 or 17.5 and 21 + 3.5 or 24.5.

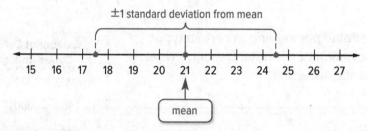

±1 standard deviation from mean

15  16  17  18  19  20  21  22  23  24  25  26  27

mean

## Example

Tutor

**2.** The standard deviation of quiz scores for Class A is about 1.2. Describe the quiz scores that are within one standard deviation of the mean.

| Quiz Scores, Class A | | | |
|---|---|---|---|
| 9 | 8 | 6 | 7 |
| 8 | 9 | 9 | 10 |
| 7 | 10 | 8 | 8 |

**Step 1** Find the mean.

$$\text{mean} = \frac{9 + 8 + \dots + 8}{12} = 8.25$$

**Step 2** Find the range of values that are within one standard deviation of the mean.

$8.25 - 1.2 = 7.05$    Subtract the standard deviation from the mean.

$8.25 + 1.2 = 9.45$    Add the standard deviation to the mean.

Quiz scores that are between 7.05 and 9.45 points are within one standard deviation of the mean.

**Got it?** Do this problem to find out.

b. _____

*Show your work.*

b. The standard deviation of quiz scores for Class B is about 1.9. Describe the quiz scores that are within one standard deviation of the mean.

| Quiz Scores, Class B | | | |
|---|---|---|---|
| 8 | 5 | 3 | 7 |
| 7 | 9 | 7 | 9 |
| 7 | 8 | 10 | 10 |

# Guided Practice

Check ✓

1. The table shows the milligrams of caffeine per serving in certain types of tea. Find the mean absolute deviation of the set of data. Describe what the mean absolute deviation represents. (Example 1)

| Amount of Caffeine in Tea (milligrams) | | | | |
|---|---|---|---|---|
| 9 | 46 | 18 | 35 | 30 |
| 12 | 56 | 24 | 38 | 32 |

_____

_____

*Show your work.*

2. The table shows the milligrams of caffeine per serving in certain types of coffee. Find the mean absolute deviation of the data. Describe what the mean absolute deviation represents. (Example 1)

| Amount of Caffeine in Coffee (milligrams) | | |
|---|---|---|
| 145 | 170 | 150 |
| 90 | 100 | 100 |
| 165 | 135 | 106 |

_____

_____

3. Refer to the table in Exercise 1. The standard deviation of the amounts of caffeine is about 14 milligrams. Describe the data values that are within one standard deviation of the mean. (Example 2)

_____

_____

4. **Building on the Essential Question** How does the mean absolute deviation describe the variation of a set of data? _____

_____

_____

_____

**Rate Yourself!**

How confident are you about measures of variation? Color the square that applies.

For more help, go online to access a Personal Tutor.

Tutor

# Independent Practice

Go online for Step-by-Step Solutions

**Find the mean absolute deviation of each set of data. Round to the nearest tenth if necessary. Describe what the mean absolute deviation represents.** (Example 1)

| Average Speeds of Selected Animals (mph) | | |
|---|---|---|
| 70 | 40 | 45 |
| 42 | 40 | 36 |

_____

_____

_____

**2.**

| Average Numbers of Annual Vacation Days for Selected Countries | | | | | | |
|---|---|---|---|---|---|---|
| 34 | 26 | 37 | 35 | 42 | 25 | 25 |

_____

_____

_____

 Refer to the table in Exercise 1. The standard deviation of the average speeds of some animals is about 11.3 miles per hour. Describe the data values that are within one standard deviation of the mean. (Example 2)

_____

_____

**4.** **MP** **Justify Conclusions** The table shows the total points scored in beach volleyball matches.

a. Find the mean absolute deviation for each set of data. Round to the nearest hundredth. Then write a few sentences comparing their variation.

_____

_____

_____

b. The standard deviation of the men's scores is 6.6 points. The standard deviation of the women's scores is 10.3 points. Describe how this information supports your answer to part **a**.

_____

_____

_____

_____

_____

| Beach Volleyball Scores | |
|---|---|
| **Men** | **Women** |
| 52 | 47 |
| 61 | 42 |
| 42 | 42 |
| 44 | 42 |
| 60 | 17 |
| 50 | 54 |
| 55 | 52 |
| 42 | 42 |
| 49 | 29 |
| 46 | 37 |

5. **MP Find the Error** Brian is describing the data values that are within one standard deviation of the mean of a set of data. Find his mistake and correct it.

_____

_____

_____

_____

> Less than half of my data values are within one standard deviation of the mean.

6. **MP Identify Structure** Create a list of data with at least five numbers that has a range of 40. Describe the mean absolute deviation.

_____

_____

7. **MP Persevere with Problems** The standard deviation of ribbon lengths is about 7.2 inches. Describe the lengths that are within two standard deviations of the mean. Explain your reasoning.

| Lengths of Ribbon (in.) | | | |
|---|---|---|---|
| 42 | 24 | 48 | 36 |
| 28 | 36 | 36 | 30 |

_____

_____

_____

8. **MP Justify Conclusions** Determine whether the following statement is *always*, *sometimes*, or *never* true. Justify your response.

*A data set with a mean absolute deviation of 9 is more spread out than a data set with a mean absolute deviation of 3.*

_____

_____

_____

9. **MP Reason Inductively** Compare and contrast standard deviation and mean absolute deviation.

_____

_____

_____

_____

_____

# Extra Practice

**Copy and Solve** **For Exercises 10–16, show your work and answers on a separate piece of paper.**

**10.** The table shows the number of hours of sleep for selected animals, rounded to the nearest hour.

| Daily Sleep | | | | | |
|---|---|---|---|---|---|
| **Animal** | brown bat | giant armadillo | cat | bottle-nosed dolphin | gray seal | horse |
| **Amount of Sleep (h)** | 20 | 18 | 12 | 10 | 6 | 3 |

   **a.** Find the mean absolute deviation of the set of data. Describe what the mean absolute deviation represents.

   **b.** The standard deviation of the data is about 6 hours. Describe the data that are within one standard deviation of the mean.

**11** The table shows the speeds of eight roller coasters in the United States.

| Roller Coaster Speeds | | | | | | | | |
|---|---|---|---|---|---|---|---|---|
| **Coaster** | Dodonpa | Kingda Ka | Millennium Force | Phantom's Revenge | Steel Dragon 2000 | Superman: The Escape | Top Thrill Dragster | Tower of Terror |
| **Speed (mph)** | 107 | 128 | 93 | 82 | 95 | 100 | 120 | 100 |

   **a.** Find the mean absolute deviation of the set of data. Round to the nearest hundredth if necessary. Describe what the mean absolute deviation represents.

   **b.** The standard deviation of the data is about 13.9 miles per hour. Describe the data that are within one standard deviation of the mean. Round to the nearest hundredth if necessary.

**12.** The table shows the bids that some comic books received at an auction.

| Comic Book Bids ($) | | | | |
|---|---|---|---|---|
| 3.25 | 4.50 | 5.00 | 5.75 | 2.25 |
| 8.50 | 6.00 | 3.50 | 4.50 | 5.00 |

   **a.** Find the mean absolute deviation of the set of data. Round to the nearest hundredth if necessary. Describe what the mean absolute deviation represents.

   **b.** The standard deviation of the data is about $1.64. Describe the data that are within one standard deviation of the mean. Round to the nearest hundredth if necessary.

**13.** The table shows the lengths of four different boards. The standard deviation of the lengths is about 2.9 feet. Determine which statements are true. Select all that apply.

| Length (ft) | | | |
|---|---|---|---|
| 12 | 15 | 15 | 20 |

☐ The mean absolute deviation is less than the standard deviation.

☐ The range is greater than the mean absolute deviation.

☐ The standard deviation is greater than the range.

**14.** The numbers of cans donated by five students during a canned food drive are shown in the table.

| Number of Cans Donated | | | | |
|---|---|---|---|---|
| 8 | 10 | 14 | 22 | 16 |

What is the mean absolute deviation of the data? ☐

## CCSS Common Core Spiral Review

**15.** The table shows lengths of rivers in two continents. **6.SP.3, 6.SP.5**

   **a.** Which continent has a greater range of length of rivers?

   **b.** Find the measures of center for each continent.

   **c.** Select the appropriate measure of center or range to describe the lengths of rivers for each continent. Justify your response.

   **d.** Find the measures of variation for each continent.

| Length (miles) of Principal Rivers | | | |
|---|---|---|---|
| Africa | | South America | |
| 4,160 | 700 | 4,000 | 1,300 |
| 2,900 | 660 | 2,485 | 1,100 |
| 2,590 | 500 | 2,100 | 1,000 |
| 1,700 | 1,100 | 2,013 | 1,000 |
| 1,300 | 1,020 | 1,988 | 1,000 |
| 1,100 | 1,000 | 1,750 | 956 |
| 1,000 | | 1,677 | 910 |
| | | 1,600 | 808 |
| | | 1,584 | 400 |
| | | 1,400 | 150 |

**16.** Find the measures of center for the set of data. Round to the nearest tenth if necessary. **6.SP.3, 6.SP.5**

**Tallest Buildings in Dallas, Texas**

| Stem | Leaf |
|---|---|
| 2 | 7 9 9 |
| 3 | 0 1 1 1 3 3 4 4 6 6 7 |
| 4 | 0 2 2 5 9 |
| 5 | 0 0 0 0 2 5 6 8 |
| 6 | 0 |
| 7 | 2 |

2|7 = 27

# Analyze Data Distributions

## Vocabulary Start-Up

Recall that in statistical displays, peaks, gaps, clusters, and outliers are identified easily.

**Complete the graphic organizer by matching the term with the correct description.**

| | |
|---|---|
| A gap is... | ...the most frequently occurring value or interval of value. |
| A peak is... | ...when many data values are grouped together. |
| An outlier is... | ...a data value that is 1.5 times the interquartile range from the first or third quartiles. |
| A cluster is... | ...where there are no data values. |

### Essential Question

HOW are patterns used when comparing two quantities?

 **Vocabulary**

distribution
symmetric

### Common Core State Standards

**Content Standards**
Preparation for S.ID.2 and S.ID.3
**MP** Mathematical Practices
1, 3, 4

##  Real-World Link

Find a photo of a mountain range. Describe clusters, gaps, and peaks in terms of the photo.

_____

_____

### Which **MP** Mathematical Practices did you use?
### Shade the circle(s) that applies.

① Persevere with Problems          ⑤ Use Math Tools

② Reason Abstractly                ⑥ Attend to Precision

③ Construct an Argument            ⑦ Make Use of Structure

④ Model with Mathematics           ⑧ Use Repeated Reasoning

# Describe a Distribution by Shape

The **distribution** of a set of data shows the arrangement of data values. It can be described by its center, spread (variation), and overall shape. Determining the symmetry of the distribution is one way to describe shape. If the left side of a distribution looks like the right side, then the distribution is **symmetric**.

**Symmetric**

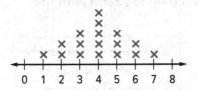

The left side looks like the right side.

**Non-Symmetric**

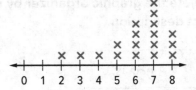

The right side is taller than the left side.

Another way to describe the shape of a distribution is to identify peaks, clusters, gaps, and outliers. If there is an outlier, the distribution is not symmetric.

## Example

**Tutor**

**1.** **The graph shows the weights of adult cats. Identify any symmetry, clusters, gaps, peaks, or outliers in the distribution.**

The distribution is non-symmetric. There is a cluster from 7—12 with a peak at 10. There is a gap between 12 and 14, and there are no outliers.

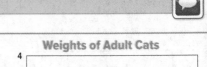
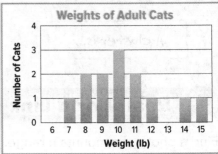

**Got it?** Do this problem to find out.

Show your work.

a. Identify any symmetry, clusters, gaps, peaks, or outliers in the distribution below.

**Height of Students (in.)**

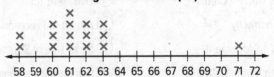

a. _____

# Describe the Center and Spread of a Distribution

The shape of a distribution tells you which measures are most appropriate for describing the center and spread of a distribution. The mean and mean absolute deviation are affected by outliers, while the median and interquartile range are resistant to outliers.

Use the following graphic organizer to decide which measures of center and spread are most appropriate to describe a data distribution.

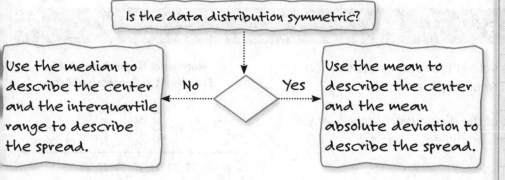

Is the data distribution symmetric?

No → Use the median to describe the center and the interquartile range to describe the spread.

Yes → Use the mean to describe the center and the mean absolute deviation to describe the spread.

## Example

Tutor

2. Mr. Watkin's class charted the high temperatures in various cities. The results are shown in the line plot.

**Temperature**

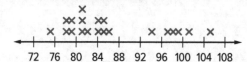

```
            ×
        ×× × ××
    ×  ×× ×× ×××     ×  ××× ×   ×
  ┼──┼──┼──┼──┼──┼──┼──┼──┼──┼──┼──
  72  76  80  84  88  92  96  100 104 108
```

**Describe the center and spread of the distribution. Justify your response based on the shape of the distribution.**

The distribution is not symmetric. So, the median and interquartile range are the appropriate measures to use. The data are centered around the median of 84°. The first quartile is 80 and the third quartile is 95.5. So, the interquartile range is 95.5 − 80 or 15.5°. The spread of the data around the center is 15.5°.

**Got it?** Do this problem to find out.

b. _____

b. The graph shows the hours per week that dance students practice their dances. Describe the center and spread of the distribution. Justify your response based on the shape of the distribution. Round to the nearest tenth if necessary.

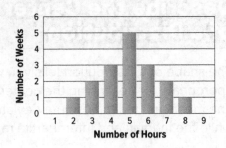

## Guided Practice

1. The number of nachos sold at the football concession stand is shown in the line plot at the right.

   a. Describe the shape of the distribution. Identify any clusters, gaps, peaks, or outliers. (Example 1)

   _____

   _____

   _____

**Number of Nacho Bowls Sold Each Night at Concession Stand**

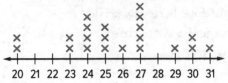

   b. Describe the center and spread of the distribution. Justify your response based on the shape of the distribution. (Example 2)

   _____

   _____

   _____

   _____

2.  **Building on the Essential Question** Why is the median used to describe the center of a non-symmetric distribution instead of the mean?

   _____

   _____

   _____

   _____

**Rate Yourself!**

How confident are you about describing data distributions? Check the box that applies.

For more help, go online to access a Personal Tutor.

Name _____ My Homework _____

# Independent Practice

Go online for Step-by-Step Solutions

**1** The scores for Ms. Hermes math class are shown in the histogram. Describe the shape of the distribution shown. Identify any clusters, gaps, peaks, or outliers. (Example 1)

_____

_____

_____

_____

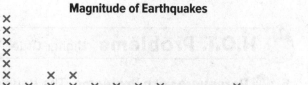

**Test Scores**

_(histogram: Number of Students vs. Percent; bars at 51–60, 61–70, 71–80, 81–90, 91–100)_

**2.** The magnitude of several earthquakes is shown in the line plot at the right.

   **a.** Describe the shape of the distribution shown. Identify any clusters, gaps, peaks, or outliers. (Example 1)

   _____

   _____

   _____

**Magnitude of Earthquakes**

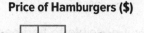

_(line plot with × marks from 1.5 to 4)_

   **b.** Describe the center and spread of the distribution. Justify your response based on the shape of the distribution. (Example 2)

   _____

   _____

   _____

**3** The box plot shows the prices of hamburgers at different restaurants.

   **a.** Describe the shape of the distribution using symmetry and outliers. (Example 1)

   _____

   _____

   _____

**Price of Hamburgers ($)**

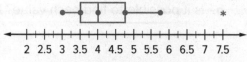

_(box plot from 2 to 7.5 with outlier at *)_

   **b.** Describe the center and spread of the distribution. Justify your response based on the shape of the distribution. (Example 2)

   _____

   _____

   _____

   _____

4. **MP** **Make a Conjecture** A distribution that is not symmetric is called *skewed*. A distribution can be skewed left or right. It is skewed left if the data are more spread out on the left side than the right side. Is the distribution shown skewed left or skewed right? Explain your reasoning to a classmate.

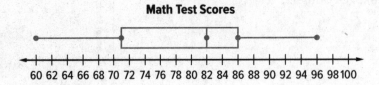

Math Test Scores

60 62 64 66 68 70 72 74 76 78 80 82 84 86 88 90 92 94 96 98 100

_____

_____

## H.O.T. Problems Higher Order Thinking

5. **MP** **Persevere with Problems** The double box plot shows scores for a football team.

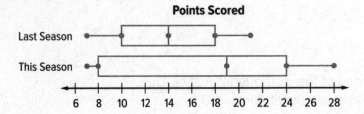

Points Scored

Last Season

This Season

6  8  10  12  14  16  18  20  22  24  26  28

a. Choose the measures that are appropriate to describe the center and spread of each box plot. Explain. _____

_____

_____

b. Is it possible to find each value? Explain. _____

_____

6. **MP** **Persevere with Problems** Explain why you cannot describe the specific location of the center and spread of the box plot shown using the most appropriate measures.

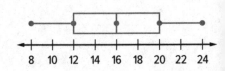

8  10  12  14  16  18  20  22  24

_____

_____

_____

_____

# Extra Practice

**Copy and Solve** For Exercises 7–17, show your work and answers on a separate piece of paper.

**7** The winning scores for twenty Super Bowls are shown in the histogram below. Describe the shape of the distribution. Identify any clusters, gaps, peaks, or outliers.

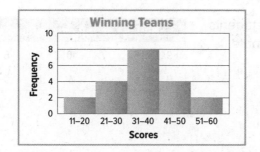

8. Wyatt took a survey of the number of times his classmates when to the movies this month. The results are shown in the line plot below.

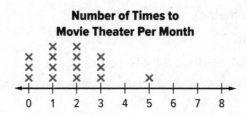

   a. Describe the shape of the distribution. Identify any clusters, gaps, peaks, or outliers.

   b. Describe the center and spread of the distribution. Justify your response based on the shape of the distribution.

9. The box plot shows the visitors to a butterfly exhibit each day for a month.

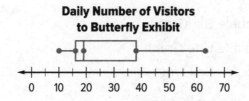

   a. Describe the shape of the distribution using symmetry and outliers.

   b. Describe the center and spread of the distribution. Justify your response based on the shape of the distribution.

10. **MP** **Justify Conclusions** Examine the data displays in Exercises 7–9. Determine if any of the distributions are skewed left or skewed right. Explain.

11. The box plot shows the number of hours spent working on a science project by students. Determine if each statement is a valid conclusion based on the box plot. Select yes or no.

**Time Working (h)**

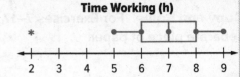

a. The distribution is symmetric. ☐ Yes ☐ No

b. The median is the best measure to describe the center. ☐ Yes ☐ No

c. There is an outlier at 2. ☐ Yes ☐ No

12. The list of data shows the number of students at different bus stops on Mr. Carter's route. Construct a line plot of the data.

| Number of Students at Bus Stops | | | | | | |
|---|---|---|---|---|---|---|
| 9 | 5 | 7 | 10 | 2 | 6 | 4 |
| 9 | 5 | 6 | 4 | 5 | 9 | 7 |

**Number of Students
at Bus Stops**

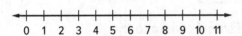

Circle the correct term to make each statement true.

a. The distribution is (symmetric, not symmetric).

b. There is a (gap, cluster) between 4 and 7.

c. The (mean, median) is the best measure to describe the center.

**CCSS** **Common Core Spiral Review**

**Select an appropriate display for each situation. Choose from the list shown.** 6.SP.4

| | |
|---|---|
| Bar Graph | Line Graph |
| Box Plot | Line Plot |
| Circle Graph | Double Bar Graph |
| Scatter Plot | Double Line Graph |
| Histogram | Double Box Plot |

13. the number of cell phone subscribers for the past 5 years

14. point totals for the top 10 NASCAR drivers

15. the portion of a family's budget assigned to each category

16. the median of the exam scores for one class

17. gas mileage for 2013 cars

# 21ST CENTURY CAREER
## in Marketing

## Sports Marketer

Are you creative and competitive? Would you enjoy a job working in the sports business? If so, you should consider a career in sports marketing. Sports marketers use statistics to develop plans to promote sporting events, such as state athletic games. They also work for professional and college sports teams, Olympic athletes, and sporting event venues. Their job is to develop merchandise and plan events that promote an athlete's or team's popularity, thereby increasing sales.

**College & Career**
**READINESS**

Explore college and careers at ccr.mcgraw-hill.com

## Is This the Career for You?

Are you interested in a career as a sports marketer? Take some of the following courses in high school.

◆ Calculus for Business
◆ Principles of Marketing
◆ Entertainment Essentials
◆ Statistical Methods

Turn the page to find out how math relates to a career in Marketing.

## MP Promoting the Games

**Use the information in the table to solve each problem.**

**1.** On the coordinate grid, graph each point. Then draw a line of best fit. What two points did you use to draw the line?

**2.** Write an equation in slope-intercept form for the line of fit.

**3.** What do the slope and *y*-intercept of the line of fit represent?

| Olympic Winning Times for 100-meter Men's Freestyle | | | |
|---|---|---|---|
| Years Since 1960 | Time (min) | Years Since 1960 | Time (min) |
| 4 | 53.4 | 8 | 52.2 |
| 12 | 51.22 | 16 | 49.99 |
| 20 | 50.40 | 24 | 49.8 |
| 28 | 48.63 | 32 | 49.02 |
| 36 | 48.74 | 40 | 48.30 |
| 44 | 48.17 | 48 | 47.21 |

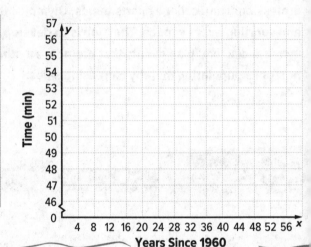

## MP Career Project

It's time to update your career portfolio! Find sports marketing blogs on the Internet and use them to answer these questions: Which sport(s) did the bloggers write about? What did you learn about sports marketing? Were there any common themes or attitudes among the sports marketing bloggers?

List the strengths you have that would help you succeed in this career.

- 
- 
- 
- 
-

# Chapter Review

## Vocabulary Check

**Complete each sentence using the vocabulary list at the beginning of the chapter. Then circle the word that completes the sentence in the word search.**

1. Data with one variable are called

   _____.

2. _____ is the ratio of the value of a subtotal to the value of the total.

3. Data with two variables are called

   _____.

4. Data that can be measured are

   _____.

5. The _____ uses numbers to characterize a set of data.

6. A _____ shows the relationship between data graphed as ordered pairs on the coordinate plane.

7. In a scatter plot, the _____ fit is close to most of the data points on the coordinate plane.

8. The arrangement of data values is called

   a _____.

9. When the left side of a distribution looks like the right side, the distribution is

   _____.

10. Data that can be observed or described

    are _____.

11. A _____ shows data that pertain to two different categories.

12. The _____ is a numerical value that shows how the data deviated from the mean. _____

13. The average distance between each data value and the mean is called the

    _____.

```
L V P R M U V B P W L R J Y G L V N Y E K S A T T
N A J N N M E A Q H G K V Y C T E B O F L C T U I
N O I T A I V E D D R A D N A T S J F O T A A N F
U G U H R D I S T R I B U T I O N L E Q J T D I T
J K B E C A D I S F S I N Q C N G T F A N T E V S
E N O R M S D T Z Y H W E G E M Z A J K P E V A E
D N T I X D W O G G S W Y O X A K A U Z I R I R B
F W B I U P P W Z U F W S V S U X F S L A P T I F
M E A N A B S O L U T E D E V I A T I O N L A A O
A A Y C N E U Q E R F E V I T A L E R L O O T T E
Q U A L I T A T I V E D A T A Z Q L F K P T I E N
H X T O Q J E D P E L B A T Y A W O W T T L T D I
A V R E R F I V E N U M B E R S U M M A R Y N A L
T J Y D E B S Y M M E T R I C H A X V J A W A T M
A T A D E T A I R A V I B B G J S P H N P Z U A D
D J Z P S F F U N P P C L S Z C L C Y Z S S Q B D
```

## Use Your FOLDABLES

**Use your Foldable to help review the chapter.**

Tape here

A Line of Best Fit is useful for:

A Two-Way Table is useful for:

A Scatter Plot is useful for:

## Got it?

**Number and perform the steps in the correct order to write an equation for the line of best fit for the scatter plot.**

_____ Write the equation in $y = mx + b$ form. _____

_____ Find the $y$-intercept. _____

_____ Draw the line. _____

_____ Choose two points. _____

_____ Find the slope. _____

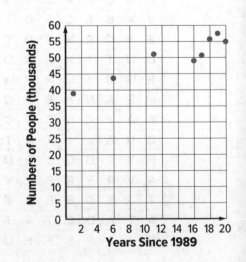

# Power Up! Performance Task

## Grand Expansion

Midtown Independent School is making plans to move to a new building to meet the increasing number of students in its middle school. The school's enrollment over the last several years is shown on the scatter plot.

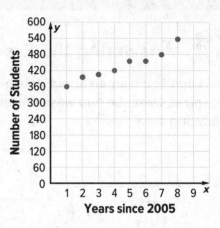

**Write your answers on another piece of paper. Show all of your work to receive full credit.**

### Part A

Draw a line of best fit on the graph. Interpret the scatter plot based on the shape of the distribution. Explain what this means in regard to enrollment.

### Part B

Write an equation for the line of best fit. What do the slope and *y*-intercept represent?

### Part C

The school must move to the new building once enrollment exceeds 690 students. If the enrollment continues to increase at the current rate, what year must the new building be ready?

### Part D

The two-way table shows the number of boys and girls who play sports in the middle school and high school. Find the relative frequencies of the students by columns to the nearest percent. Two hundred fifty new lockers will be purchased. The new lockers will be placed in the locker rooms based on the current distribution of students playing sports. Determine how the lockers should be distributed. Explain.

|  | Boys | Girls | Total |
|---|---|---|---|
| Middle School | 40 | 35 | 75 |
| High School | 30 | 45 | 75 |
| Total | 70 | 80 | 150 |

# Reflect

Use what you learned about data analysis to complete the graphic organizer. Describe two ways patterns are used in each concept when analyzing data.

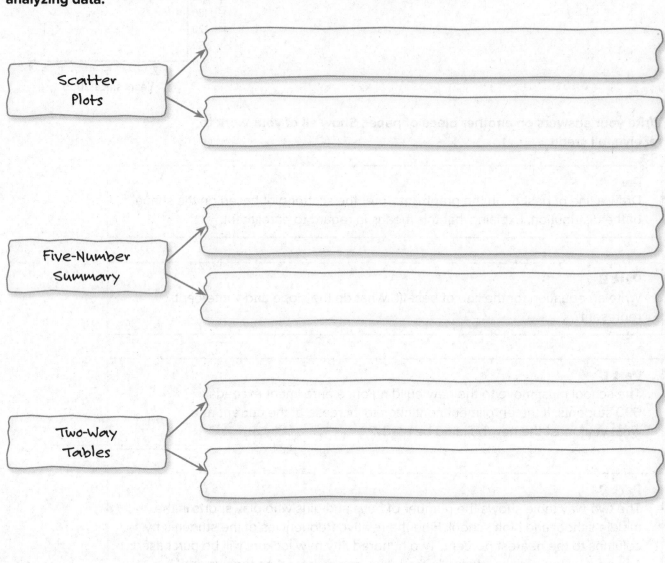

Scatter Plots

Five-Number Summary

Two-Way Tables

 **Answer the Essential Question.** HOW are patterns used when comparing two quantities?

_____

_____

_____

_____

# UNIT PROJECT

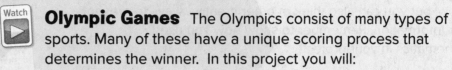

**Olympic Games** The Olympics consist of many types of sports. Many of these have a unique scoring process that determines the winner. In this project you will:

- **Collaborate** with your classmates as you gather Olympics statistics.
- **Share** the results of your research in a creative way.
- **Ⓔ Reflect** on why learning mathematics is important.

By the end of this project, you will understand how scatter plots and data analysis are involved in presenting Olympic Games statistics.

**⏻ Go Online** Work with your group to research and complete each activity. You will use your results in the Share section on the following page.

1. Choose a country that has participated in basketball in the Summer Olympics. Use the Internet to research the team. Find their average points per game over the past 10 Summer Olympic Games. Record the information in a table.

2. Make a scatter plot of the data from Exercise 1. Determine if the data can be used to predict the average number of points in the next Summer Olympics. If so, make a prediction.

3. Research the number of Olympic records the U.S. has received in the Olympic sport of your choice. Use a graph of your choice and interpret the graph.

4. Research the winning scores in archery over the past 10 Summer Olympics. Draw a histogram to display the data. Interpret the graph.

5. During the ranking round in archery each player will shoot a total of 72 arrows. Create a score card for one player in the first round. Summarize the data in a box plot and interpret the graph.

Collaborate

With your group, decide on a way to share what you have learned about Olympic scoring. Some suggestions are listed below, but you can also think of other creative ways to present your information. Remember to show how you used mathematics in your project!

- Act as a television reporter for the Olympics and describe the scores and medals won in a few events. Include graphics that would appear on screen.
- Choose an Olympic sport that you do not know much about. Explain the scoring system in your sport. Then create tables and graphs to present real data from your sport in the most recent Olympics.

Check out the note on the right to connect this project with other subjects.

**connect** with Physical Education

**Global Awareness** Research information on how to play or participate in a sport that is popular in another country. Some questions to consider are:

- What are the basic rules?
- What are the jobs of the offense and the defense?

 **Reflect**

On Your Own

6. **Ⓔ Answer the Essential Question** Why is learning mathematics important?

   a. How did you use what you learned about scatter plots in this chapter to represent mathematical ideas in this project?

   _____

   _____

   _____

   _____

   b. How did you use what you learned about data analysis to communicate mathematical ideas effectively in this project?

   _____

   _____

   _____

   _____

# Glossary/Glosario

Go online for the eGlossary.

The eGlossary contains words and definitions in the following 13 languages:

| | | | | |
|---|---|---|---|---|
| Arabic | Cantonese | Hmong | Spanish | Urdu |
| Bengali | English | Korean | Tagalog | Vietnamese |
| Brazilian Portuguese | Haitian Creole | Russian | | |

## English | ## Español

### Aa

**accuracy** The degree of closeness of a measurement to the true value.

**acute angle** An angle whose measure is less than 90°.

**acute triangle** A triangle with all acute angles.

**Addition Property of Equality** If you add the same number to each side of an equation, the two sides remain equal.

**adjacent angles** Angles that share a common vertex, a common side, and do not overlap. In the figure, the adjacent angles are ∠5 and ∠6.

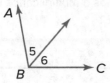

**algebra** A branch of mathematics that involves expressions with variables.

**algebraic expression** A combination of variables, numbers, and at least one operation.

**exactitud** Cercanía de una medida a su valor verdadero.

**ángulo agudo** Ángulo que mide menos de 90°.

**triángulo acutángulo** Triángulo con todos los ángulos agudos.

**propiedad de adición de la igualdad** Si sumas el mismo número a ambos lados de una ecuación, los dos lados permanecen iguales.

**ángulos adyacentes** Ángulos que comparten un vértice, un lado común y no se traslapan. En la figura, los ángulos adyacentes son ∠5 y ∠6.

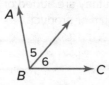

**álgebra** Rama de las matemáticas que trabaja con expresiones con variables.

**expresión algebraica** Una combinación de variables, números y por lo menos una operación.

**alternate exterior angles** Exterior angles that lie on opposite sides of the transversal. In the figure, transversal *t* intersects lines $\ell$ and *m*. $\angle 1$ and $\angle 7$, and $\angle 2$ and $\angle 8$ are alternate exterior angles. If line $\ell$ and *m* are parallel, then these pairs of angles are congruent.

**ángulos alternos externos** Ángulos externos que se encuentran en lados opuestos de la transversal. En la figura, la transversal *t* interseca las rectas $\ell$ y *m*. $\angle 1$ y $\angle 7$, y $\angle 2$ y $\angle 8$ son ángulos alternos externos. Si las rectas $\ell$ y *m* son paralelas, entonces estos ángulos son pares de ángulos congruentes.

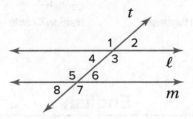

**alternate interior angles** Interior angles that lie on opposite sides of the transversal. In the figure below, transversal *t* intersects lines $\ell$ and *m*. $\angle 3$ and $\angle 5$, and $\angle 4$ and $\angle 6$ are alternate interior angles. If lines $\ell$ and *m* are parallel, then these pairs of angles are congruent.

**ángulos alternos internos** Ángulos internos que se encuentran en lados opuestos de la transversal. En la figura, la transversal *t* interseca las rectas $\ell$ y *m*. $\angle 3$ y $\angle 5$, y $\angle 4$ y $\angle 6$ son ángulos alternos internos. Si las rectas $\ell$ y *m* son paralelas, entonces estos ángulos son pares de ángulos congruentes.

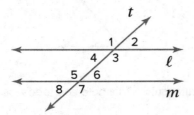

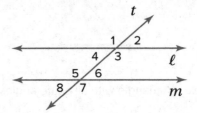

**angle of rotation** The degree measure of the angle through which a figure is rotated.

**ángulo de rotación** Medida en grados del ángulo sobre el cual se rota una figura.

**arc** One of two parts of a circle separated by a central angle.

**arco** Una de dos partes de un círculo separadas por un ángulo central.

**Associative Property** The way in which three numbers are grouped when they are added or multiplied does not change their sum or product.

**propiedad asociativa** La forma en que se agrupan tres números al sumarlos o multiplicarlos no altera su suma o producto.

---

### Bb

**base** In a power, the number that is the common factor. In $10^3$, the base is 10. That is, $10^3 = 10 \times 10 \times 10$.

**base** En una potencia, número que es el factor común. En $10^3$, la base es 10. Es decir, $10^3 = 10 \times 10 \times 10$.

**base** One of the two parallel congruent faces of a prism.

**base** Una de las dos caras paralelas congruentes de un prisma.

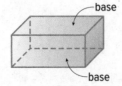

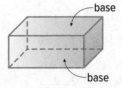

**biased sample**   A sample drawn in such a way that one or more parts of the population are favored over others.

**muestra sesgada**   Muestra en que se favorece una o más partes de una población.

**bivariate data**   Data with two variables, or pairs of numerical observations.

**datos bivariantes**   Datos con dos variables, o pares de observaciones numéricas.

**box plot**   A method of visually displaying a distribution of data values by using the median, quartiles, and extremes of the data set. A box shows the middle 50% of the data.

**diagrama de caja**   Un método de mostrar visualmente una distribución de valores usando la mediana, cuartiles y extremos del conjunto de datos. Una caja muestra el 50% del medio de los datos.

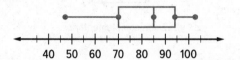

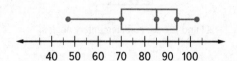

---

**center**   The given point from which all points on a circle are the same distance.

**centro**   Un punto dado del cual equidistan todos los puntos de un círculo.

**center of dilation**   The center point from which dilations are performed.

**centro de la homotecia**   Punto fijo en torno al cual se realizan las homotecias.

**center of rotation**   A fixed point around which shapes move in a circular motion to a new position.

**centro de rotación**   Punto fijo alrededor del cual se giran las figuras en movimiento circular alrededor de un punto fijo.

**central angle**   An angle that intersects a circle in two points and has its vertex at the center of the circle.

**ángulo central**   Ángulo que interseca un círculo en dos puntos y cuyo vértice es el centro del círculo.

**circle**   The set of all points in a plane that are the same distance from a given point called the center.

**círculo**   Conjunto de todos los puntos en un plano que equidistan de un punto dado llamado centro.

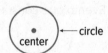

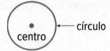

**circumference**   The distance around a circle.

**circunferencia**   La distancia alrededor de un círculo.

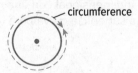

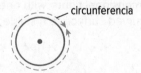

**chord**   A segment with endpoints that are on a circle.

**cuerda**   Segmento cuyos extremos están sobre un círculo.

**coefficient**   The numerical factor of a term that contains a variable.

**coeficiente**   Factor numérico de un término que contiene una variable.

**common difference**   The difference between any two consecutive terms in an arithmetic sequence.

**diferencia común**   La diferencia entre cualquier par de términos consecutivos en una sucesión aritmética.

**Commutative Property** The order in which two numbers are added or multiplied does not change their sum or product.

**propiedad conmutativa** La forma en que se suman o multiplican dos números no altera su suma o producto.

**complementary angles** Two angles are complementary if the sum of their measures is 90°.

**ángulos complementarios** Dos ángulos son complementarios si la suma de sus medidas es 90°.

∠1 and ∠2 are complementary angles.

∠1 y ∠2 son complementarios.

**composite figure** A figure that is made up of two or more shapes.

**figura compleja** Figura compuesta de dos o más formas.

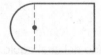

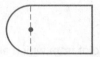

**composite solid** An object made up of more than one type of solid.

**sólido complejo** Cuerpo compuesto de más de un tipo de sólido.

**composition of transformations** The resulting transformation when a transformation is applied to a figure and then another transformation is applied to its image.

**composición de transformaciones** Transformación que resulta cuando se aplica una transformación a una figura y luego se le aplica otra transformación a su imagen.

**compound event** An event that consists of two or more simple events.

**evento compuesto** Evento que consta de dos o más eventos simples.

**compound interest** Interest paid on the initial principal and on interest earned in the past.

**interés compuesto** Interés que se paga por el capital inicial y sobre el interés ganado en el pasado.

**cone** A three-dimensional figure with one circular base connected by a curved surface to a single vertex.

**cono** Una figura tridimensional con una circular base conectada por una superficie curva para un solo vértice.

vertex

vértice

**congruent** Having the same measure; if one image can be obtained by another by a sequence of rotations, reflections, or translations.

**congruente** Que tienen la misma medida; si una imagen puede obtenerse de otra por una secuencia de rotaciones, reflexiones o traslaciones.

**constant** A term without a variable.

**constante** Término sin variables.

**constant of proportionality**   The constant ratio in a proportional linear relationship.

**constant of variation**   A constant ratio in a direct variation.

**constant rate of change**   The rate of change between any two points in a linear relationship is the same or *constant*.

**continuous data**   Data that can take on any value. There is no space between data values for a given domain. Graphs are represented by solid lines.

**convenience sample**   A sample which includes members of the population that are easily accessed.

**converse**   The converse of a theorem is formed when the parts of the theorem are reversed. The converse of the Pythagorean Theorem can be used to test whether a triangle is a right triangle. If the sides of the triangle have lengths $a$, $b$, and $c$, such that $c^2 = a^2 + b^2$, then the triangle is a right triangle.

**coordinate plane**   A coordinate system in which a horizontal number line and a vertical number line intersect at their zero points.

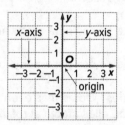

**coplanar**   Lines that lie in the same plane.

**corresponding angles**   Angles that are in the same position on two parallel lines in relation to a transversal.

**corresponding parts**   Parts of congruent or similar figures that match.

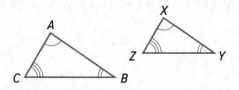

**constante de proporcionalidad**   La razón constante en una relación lineal proporcional.

**constante de variación**   Razón constante en una relación de variación directa.

**tasa constante de cambio**   La tasa de cambio entre dos puntos cualesquiera en una relación lineal permanece igual o *constante*.

**datos continuos**   Datos que pueden tomar cualquier valor. No hay espacio entre los valores de los datos para un dominio dado. Las gráficas se representan con rectas sólidas.

**muestra de conveniencia**   Muestra que incluye miembros de una población fácilmente accesibles.

**recíproco**   El recíproco de un teorema se forma cuando se invierten las partes del teorema. El recíproco del teorema de Pitágoras puede usarse para averiguar si un triángulo es un triángulo rectángulo. Si las longitudes de los lados de un triángulo son $a$, $b$ y $c$, tales que $c^2 = a^2 + b^2$, entonces el triángulo es un triángulo rectángulo.

**plano de coordenadas**   Sistema de coordenadas en que una recta numérica horizontal y una recta numérica vertical se intersecan en sus puntos cero.

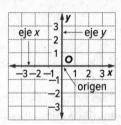

**coplanario**   Rectas que yacen en el mismo plano.

**ángulos correspondientes**   Ángulos que están en la misma posición sobre dos rectas paralelas en relación con la transversal.

**partes correspondientes**   Partes de figuras congruentes o semejantes que coinciden.

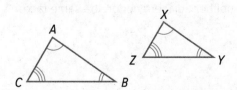

**counterexample** A statement or example that shows a conjecture is false.

**contraejemplo** Ejemplo o enunciado que demuestra que una conjetura es falsa.

**cross section** The intersection of a solid and a plane.

**sección transversal** Intersección de un sólido y un plano.

**cube root** One of three equal factors of a number. If $a^3 = b$, then $a$ is the cube root of $b$. The cube root of 64 is 4 since $4^3 = 64$.

**raíz cúbica** Uno de tres factores iguales de un número. Si $a^3 = b$, entonces $a$ es la raíz cúbica de $b$. La raíz cúbica de 64 es 4, dado que $4^3 = 64$.

**cylinder** A three-dimensional figure with two parallel congruent circular bases connected by a curved surface.

**cilindro** Una figura tridimensional con dos paralelas congruentes circulares bases conectados por una superficie curva.

**deductive reasoning** A system of reasoning that uses facts, rules, definitions, or properties to reach logical conclusions.

**razonamiento deductivo** Sistema de razonamiento que emplea hechos, reglas, definiciones o propiedades para obtener conclusions lógicas.

**defining a variable** Choosing a variable and a quantity for the variable to represent in an expression or equation.

**definir una variable** El elegir una variable y una cantidad que esté representada por la variable en una expresión o en una ecuación.

**degree** A unit used to measure angles.

**grado** Unidad que se usa para medir ángulos.

**degree** A unit used to measure temperature.

**grado** Unidad que se usa para medir la temperatura.

**dependent events** Two or more events in which the outcome of one event does affect the outcome of the other event or events.

**eventos dependientes** Dos o más eventos en que el resultado de uno de ellos afecta el resultado de los otros eventos.

**dependent variable** The variable in a relation with a value that depends on the value of the independent variable.

**variable dependiente** La variable en una relación cuyo valor depende del valor de la variable independiente.

**derived unit** A unit that is derived from a measurement system base unit, such as length, mass, or time.

**unidad derivada** Unidad derivada de una unidad básica de un sistema de medidas como por ejemplo, la longitud, la masa o el tiempo.

**diagonal** A line segment whose endpoints are vertices that are neither adjacent nor on the same face.

**diagonal** Segmento de recta cuyos extremos son vértices que no son ni adyacentes ni yacen en la misma cara.

**diameter**   The distance across a circle through its center.

diameter

**diámetro**   La distancia a través de un círculo pasando por el centro.

diámetro

**dilation**   A transformation that enlarges or reduces a figure by a scale factor.

**homotecia**   Transformación que produce la ampliación o reducción de una imagen por un factor de escala.

**dimensional analysis**   The process of including units of measurement when you compute.

**análisis dimensional**   Proceso que incorpora las unidades de medida al hacer cálculos.

**direct variation**   A relationship between two variable quantities with a constant ratio.

**variación directa**   Relación entre dos cantidades variables con una razón constante.

**discount**   The amount by which a regular price is reduced.

**descuento**   La cantidad de reducción del precio normal.

**discrete data**   Data with space between possible data values. Graphs are represented by dots.

**datos discretos**   Datos con espacios entre posibles valores de datos. Las gráficas están representadas por puntos.

**disjoint events**   Events that cannot happen at the same time.

**eventos disjuntos**   Eventos que no pueden ocurrir al mismo tiempo.

**Distance Formula**   The distance $d$ between two points with coordinates $(x_1, y_1)$ and $(x_2, y_2)$ is given by the formula

$$d = \sqrt{(x_1 - x_2)^2 + (y_1 - y_2)^2}.$$

**fórmula de la distancia**   La distancia $d$ entre dos puntos con coordenadas $(x_1, y_1)$ and $(x_2, y_2)$ viene dada por la fórmula

$$d = \sqrt{(x_1 - x_2)^2 + (y_1 - y_2)^2}.$$

**distribution**   A way to show the arrangement of data values.

**distribución**   Una manera de mostrar la agrupación de valores.

**Distributive Property**   To multiply a sum by a number, multiply each addend by the number outside the parentheses.

$$5(x + 3) = 5x + 15$$

**propiedad distributiva**   Para multiplicar una suma por un número, multiplica cada sumando por el número fuera de los paréntesis.

$$5(x + 3) = 5x + 15$$

**Division Property of Equality**   If you divide each side of an equation by the same nonzero number, the two sides remain equal.

**propiedad de división de la igualdad**   Si cada lado de una ecuación se divide entre el mismo número no nulo, los dos lados permanecen iguales.

**domain**   The set of $x$-coordinates in a relation.

**dominio**   Conjunto de coordenadas $x$ en una relación.

**double box plot**   Two box plots graphed on the same number line.

**doble diagrama de puntos**   Dos diagramas de caja sobre la misma recta numérica.

## Ee

**edge**   The line segment where two faces of a polyhedron intersect.

**arista**   El segmento de línea donde se cruzan dos caras de un poliedro.

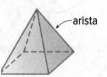

**equation**   A mathematical sentence stating that two quantities are equal.

**ecuación**   Enunciado matemático que establece que dos cantidades son iguales.

**equiangular**   A polygon in which all angles are congruent.

**equiangular**   Polígono en el cual todos los ángulos son congruentes.

**equilateral triangle**   A triangle with three congruent sides.

**triángulo equilátero**   Triángulo con tres lados congruentes.

**equivalent expressions**   Expressions that have the same value regardless of the value(s) of the variable(s).

**expresiones equivalentes**   Expresiones que poseen el mismo valor, sin importar los valores de la(s) variable(s).

**event**   An outcome is a possible result.

**evento**   Un resultado posible.

**experimental probability**   An estimated probability based on the relative frequency of positive outcomes occurring during an experiment.

**probabilidad experimental**   Probabilidad estimada que se basa en la frecuencia relativa de los resultados positivos que ocurren durante un experimento.

**exponent**   In a power, the number of times the base is used as a factor. In $10^3$, the exponent is 3.

**exponente**   En una potencia, el número de veces que la base se usa como factor. En $10^3$, el exponente es 3.

**exponential function**   A nonlinear function in which the base is a constant and the exponent is an independent variable.

**función exponencial**   Función no lineal en la cual la base es una constante y el exponente es una variable independiente.

**exterior angles**   The four outer angles formed by two lines cut by a transversal.

**ángulo externo**   Los cuatro ángulos exteriores que se forman cuando una transversal corta dos rectas.

## Ff

**face**   A flat surface of a polyhedron.

**cara**   Una superficie plana de un poliedro.

**fair game**   A game where each player has an equally likely chance of winning.

**juego justo**   Juego donde cada jugador tiene igual posibilidad de ganar.

**five-number summary**   A way of characterizing a set of data that includes the minimum, first quartile, median, third quartile, and the maximum.

**resumen de los cinco números**   Una manera de caracterizar un conjunto de datos que incluye el mínimo, el primer cuartil, la mediana, el tercer cuartil y el máximo.

**formal proof**   A two-column proof containing statements and reasons.

**demonstración formal**   Demonstración endos columnas contiene enunciados y razonamientos.

**function**   A relation in which each member of the domain (input value) is paired with exactly one member of the range (output value).

**función**   Relación en la cual a cada elemento del dominio (valor de entrada) le corresponde exactamente un único elemento del rango (valor de salida).

**function table**   A table organizing the domain, rule, and range of a function.

**tabla de funciones**   Tabla que organiza la regla de entrada y de salida de una función.

**Fundamental Counting Principle**   Uses multiplication of the number of ways each event in an experiment can occur to find the number of possible outcomes in a sample space.

**principio fundamental de contar**   Método que usa la multiplicación del número de maneras en que cada evento puede ocurrir en un experimento, para calcular el número de resultados posibles en un espacio muestral.

## Gg

**geometric sequence**   A sequence in which each term after the first is found by multiplying the previous term by a constant.

**sucesión geométrica**   Sucesión en la cual cada término después del primero se determina multiplicando el término anterior por una constante.

## Hh

**half-plane**   The part of the coordinate plane on one side of the boundary.

**semiplano**   Parte del plano de coordenadas en un lado de la frontera.

**hemisphere**   One of two congruent halves of a sphere.

**hemisferio**   Una de dos mitades congruentes de una esfera.

**hypotenuse**   The side opposite the right angle in a right triangle.

**hipotenusa**   El lado opuesto al ángulo recto de un triángulo rectángulo.

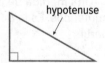
hypotenuse

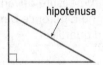
hipotenusa

## Ii

**identity**   An equation that is true for every value for the variable.

**identidad**   Ecuación que es verdad para cada valor de la variable.

**image** The resulting figure after a transformation.

**imagen** Figura que resulta después de una transformación.

**independent events** Two or more events in which the outcome of one event does not affect the outcome of the other event(s).

**eventos independientes** Dos o más eventos en los cuales el resultado de un evento no afecta el resultado de los otros eventos.

**independent variable** The variable in a function with a value that is subject to choice.

**variable independiente** Variable en una función cuyo valor está sujeto a elección.

**indirect measurement** A technique using properties of similar polygons to find distances or lengths that are difficult to measure directly.

**medición indirecta** Técnica que usa las propiedades de polígonos semejantes para calcular distancias o longitudes difíciles de medir directamente.

**inductive reasoning** Reasoning that uses a number of specific examples to arrive at a plausible generalization or prediction. Conclusions arrived at by inductive reasoning lack the logical certainty of those arrived at by deductive reasoning.

**razonamiento inductivo** Razonamiento que usa varios ejemplos especificos para lograr una generalización o una predicción plausible. Las conclusions obtenidas por razonamiento inductivo carecen de la certeza lógica de aquellas obtenidas por razonamiento deductivo.

**inequality** A mathematical sentence that contains <, >, ≠, ≤, or ≥.

**desigualdad** Enunciado matemático que contiene <, >, ≠, ≤, o ≥.

**inscribed angle** An angle that has its vertex on the circle. Its sides contain chords of the circle.

**ángulo inscrito** Ángulo cuyo vértice está en el círculo y cuyos lados contienen cuerdas del círculo.

**informal proof** A paragraph proof.

**demonstración informal** Demonstración en forma de párrafo.

**interest** The amount of money paid or earned for the use of money.

**interés** Cantidad que se cobra o se paga por el uso del dinero.

**interior angle** An angle inside a polygon.

**ángulo interno** Ángulo dentro de un polígono.

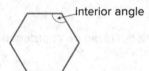
interior angle

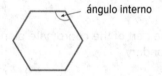
ángulo interno

**interior angles** The four inside angles formed by two lines cut by a transversal.

**ángulo interno** Los cuatro ángulos internos formados por dos rectas intersecadas por una transversal.

**interquartile range** A measure of variation in a set of numerical data. It is the difference between the first quartile and the third quartile.

**rango intercuartílico** Una medida de la variación en un conjunto de datos numéricos. Es la diferencia entre el primer y el tercer cuartil.

**inverse operations** Pairs of operations that undo each other. Addition and subtraction are inverse operations. Multiplication and division are inverse operations.

**peraciones inversas** Pares de operaciones que se anulan mutuamente. La adición y la sustracción son operaciones inversas. La multiplicación y la división son operaciones inversas.

**irrational number** A number that cannot be expressed as the quotient $\frac{a}{b}$, where $a$ and $b$ are integers and $b \neq 0$.

**números irracionales** Número que no se puede expresar como el cociente $\frac{a}{b}$, donde $a$ y $b$ son enteros y $b \neq 0$.

**isosceles triangle** A triangle with at least two congruent sides.

**triángulo isóceles** Triángulo con por lo menos dos lados congruentes.

Ll

**lateral area** The sum of the areas of the lateral faces of a solid.

**área lateral** La suma de las áreas de las caras laterales de un sólido.

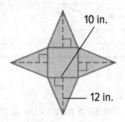

10 in.

12 in.

lateral area = $4\left(\frac{1}{2} \times 10 \times 12\right)$ = 240 square inches

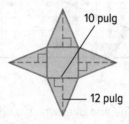

10 pulg

12 pulg

área lateral = $4\left(\frac{1}{2} \times 10 \times 12\right)$ = 240 pulgadas cuadradas

**lateral face** Any flat surface that is not a base.

**cara lateral** Cualquier superficie plana que no es la base.

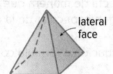

lateral face

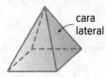

cara lateral

**legs** The two sides of a right triangle that form the right angle.

**catetos** Los dos lados de un triángulo rectángulo que forman el ángulo recto.

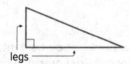

legs

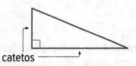

catetos

**like fractions** Fractions that have the same denominators.

**fracciones semejantes** Fracciones que tienen el mismo denominador.

**like terms** Terms that contain the same variable(s) to the same powers.

**términos semejantes** Términos que contienen la misma variable o variables elevadas a la misma potencia.

**linear** To fall in a straight line.

**lineal** Que cae en una línea recta.

**linear equation** An equation with a graph that is a straight line.

**ecuación lineal** Ecuación cuya gráfica es una recta.

**linear function** A function in which the graph of the solutions forms a line.

**función lineal** Función en la cual la gráfica de las soluciones forma un recta.

**linear relationship**  A relationship that has a straight-line graph.

**relación lineal**  Relación cuya gráfica es una recta.

**line of best fit**  A line that is very close to most of the data points in a scatter plot.

**recta de mejor ajuste**  Recta que más se acerca a la mayoría de puntos de los datos en un diagrama de dispersión.

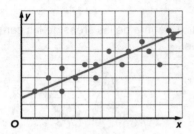

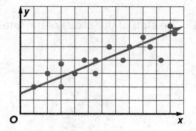

**line of reflection**  The line over which a figure is reflected.

**línea de reflexión**  Línea a través de la cual se refleja una figura.

**line of symmetry**  Each half of a figure is a mirror image of the other half when a line of symmetry is drawn.

**eje de simetría**  Recta que divide una figura en dos mitades especulares.

**line symmetry**  A figure has line symmetry if a line can be drawn so that one half of the figure is a mirror image of the other half.

**simetría lineal**  Una figura tiene simetría lineal si se puede trazar una recta de manera que una mitad de la figura sea una imagen especular de la otra mitad.

**literal equation**  An equation or formula that has more than one variable.

**ecuación literal**  Ecuación o fórmula con más de una variable.

---

## Mm

**markup**  The amount the price of an item is increased above the price the store paid for the item.

**margen de utilidad**  Cantidad de aumento en el precio de un artículo por encima del precio que paga la tienda por dicho artículo.

**mean**  The sum of the data divided by the number of items in the set.

**media**  La suma de datos dividida entre el número total de artículos.

**mean absolute deviation**  The average of the absolute values of differences between the mean and each value in a data set.

**desviación media absoluta**  El promedio de los valores absolutos de diferencias entre el medio y cada valor de un conjunto de datos.

**measures of center**  Numbers that are used to describe the center of a set of data.These measures include the mean, median, and mode.

**medidas del centro**  Números que describen el centro de un conjunto de datos. Estas medidas incluyen la media, la mediana y la moda.

**measures of variation**  Numbers used to describe the distribution or spread of a set of data.

**medidas de variación**  Números que se usan para describir la distribución o separación de un conjunto de datos.

**median**   A measure of center in a set of numerical data. The median of a list of values is the value appearing at the center of a sorted version of the list—or the mean of the two central values, if the list contains an even number of values.

**mediana**   Una medida del centro en un conjunto de datos numéricos. La mediana de una lista de valores es el valor que aparece en el centro de una versión ordenada de la lista, o la media de los dos valores centrales si la lista contiene un número par de valores.

**mode**   The number(s) or item(s) that appear most often in a set of data.

**moda**   El número(s) o artículo(s) que aparece con más frecuencia en un conjunto de datos.

**monomial**   A number, a variable, or a product of a number and one or more variables.

**monomio**   Un número, una variable o el producto de un número por una o más variables.

**Multiplication Property of Equality**   If you multiply each side of an equation by the same number, the two sides remain equal.

**propiedad de multiplicación de la igualdad**   Si cada lado de una ecuación se multiplica por el mismo número, los lados permanecen iguales.

**multiplicative inverses**   Two numbers with a product of 1. The multiplicative inverse of $\frac{2}{3}$ is $\frac{3}{2}$.

**inversos multiplicativo**   Dos números cuyo producto es 1. El inverso multiplicativo de $\frac{2}{3}$ es $\frac{3}{2}$.

---

**net**   A two-dimensional pattern of a three-dimensional figure.

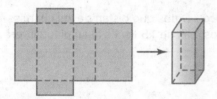

**red**   Patrón bidimensional de una figura tridimensional.

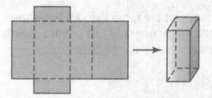

**nonlinear function**   A function whose rate of change is not constant. The graph of a nonlinear function is not a straight line.

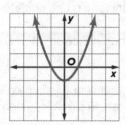

**función no lineal**   Función cuya tasa de cambio no es constante. La gráfica de una función no lineal no es una recta.

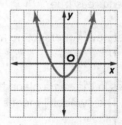

**null set**   The empty set.

**conjunto nulo**   El conjunto vacío.

---

**obtuse angle**   An angle whose measure is between 90° and 180°.

**ángulo obtuso**   Ángulo cuya medida está entre 90° y 180°.

**obtuse triangle**   A triangle with one obtuse angle.

**triángulo obtusángulo**   Triángulo con un ángulo obtuso.

**ordered pair** A pair of numbers used to locate a point in the coordinate plane. The ordered pair is written in this form: (*x*-coordinate, *y*-coordinate).

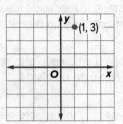

**par ordenado** Par de números que se utiliza para ubicar un punto en un plano de coordenadas. Se escribe de la siguiente forma: (coordenada *x*, coordenada *y*).

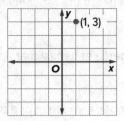

**origin** The point of intersection of the *x*-axis and *y*-axis in a coordinate plane.

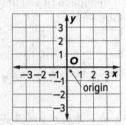

**origen** Punto en que el eje *x* y el eje *y* se intersecan en un plano de coordenadas.

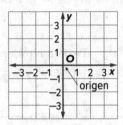

**outcome** One possible result of a probability event. For example, 4 is an outcome when a number cube is rolled.

**resultado** Una consecuencia posible de un evento de probabilidad. Por ejemplo, 4 es un resultado posible al lanzar un cubo numérico.

**outlier** Data that are more than 1.5 times the interquartile range from the first or third quartiles.

**valor atípico** Datos que distan de los cuartiles respectivos más de 1.5 veces la amplitud intercuartílica.

**Pp**

**paragraph proof** A paragraph that explains why a statement or conjecture is true.

**prueba por párrafo** Párrafo que explica por qué es verdadero un enunciado o una conjetura.

**parallel** Lines that never intersect no matter how far they extend.

**paralelo** Rectas que nunca se intersecan sea cual sea su extensión.

**parallel lines** Lines in the same plane that never intersect or cross. The symbol ∥ means parallel.

**rectas paralelas** Rectas que yacen en un mismo plano y que no se intersecan. El símbolo ∥ significa paralela a.

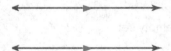

**parallelogram** A quadrilateral with both pairs of opposite sides parallel and congruent.

**paralelogramo** Cuadrilátero con ambos pares de lados opuestos, paralelos y congruentes.

**percent equation**   An equivalent form of a percent proportion in which the percent is written as a decimal.

$$\text{part} = \text{percent} \cdot \text{whole}$$

**percent of change**   A ratio that compares the change in quantity to the original amount.

$$\text{percent of change} = \frac{\text{amount of change}}{\text{original amount}}$$

**percent of decrease**   When the percent of change is negative.

**percent of increase**   When the percent of change is positive.

**percent proportion**   Compares part of a quantity to the whole quantity using a percent.

$$\frac{\text{part}}{\text{whole}} = \frac{\text{percent}}{100}$$

**perfect cube**   A rational number whose cube root is a whole number. 27 is a perfect cube because its cube root is 3.

**perfect square**   A rational number whose square root is a whole number. 25 is a perfect square because its square root is 5.

**permutation**   An arrangement or listing in which order is important.

**perpendicular lines**   Two lines that intersect to form right angles.

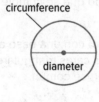

**pi**   The ratio of the circumference of a circle to its diameter. The Greek letter $\pi$ represents this number. The value of pi is always 3.1415926... .

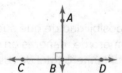

$$\pi = \frac{c}{d}$$

**point-slope form**   An equation of the form $y - y_1 = m(x - x_1)$, where $m$ is the slope and $(x_1, y_1)$ is a given point on a nonvertical line.

**ecuación porcentual**   Forma equivalente de proporción porcentual en la cual el por ciento se escribe como un decimal.

$$\text{parte} = \text{por ciento} \cdot \text{entero}$$

**porcentaje de cambio**   Razón que compara el cambio en una cantidad a la cantidad original.

$$\text{procentaje de cambio} = \frac{\text{cantidad de cambio}}{\text{cantidad original}}$$

**porcentaje de disminución**   Cuando el porcentaje de cambio es negativo.

**porcentaje de aumento**   Cuando el porcentaje de cambio es positivo.

**proporción porcentual**   Compara parte de una cantidad con la cantidad total mediante un por ciento.

$$\frac{\text{parte}}{\text{entero}} = \frac{\text{por ciento}}{100}$$

**cubo perfecto**   Número racional cuya raíz cúbica es un número entero. 27 es un cubo perfecto porque su raíz cúbica es 3.

**cuadrados perfectos**   Número racional cuya raíz cuadrada es un número entero. 25 es un cuadrado perfecto porque su raíz cuadrada es 5.

**permutación**   Arreglo o lista donde el orden es importante.

**rectas perpendiculares**   Dos rectas que se intersecan formando ángulos rectos.

**pi**   Razón de la circunferencia de un círculo al diámetro del mismo. La letra griega $\pi$ representa este número. El valor de pi es siempre 3.1415926... .

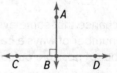

$$\pi = \frac{c}{d}$$

**forma punto-pendiente**   Ecuación de la forma $y - y_1 = m(x - x_1)$ donde m es la pendiente y $(x_1 - y_1)$ es un punto dado de una recta no vertical.

**polygon** A simple, closed figure formed by three or more line segments.

**polígono** Figura simple y cerrada formada por tres o más segmentos de recta.

**polyhedron** A three-dimensional figure with faces that are polygons.

**poliedro** Una figura tridimensional con caras que son polígonos.

**power** A product of repeated factors using an exponent and a base. The power $7^3$ is read *seven to the third power,* or *seven cubed.*

**potencia** Producto de factores repetidos con un exponente y una base. La potencia $7^3$ se lee *siete a la tercera potencia o siete al cubo.*

**precision** The ability of a measurement to be consistently reproduced.

**precisión** Capacidad de una medida a ser reproducida consistentemente.

**preimage** The original figure before a transformation.

**preimagen** Figura original antes de una transformación.

**principal** The amount of money invested or borrowed.

**capital** Cantidad de dinero que se invierte o que se toma prestada.

**prism** A polyhedron with two parallel congruent faces called bases.

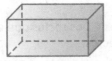

**prisma** Poliedro con dos caras congruentes y paralelas llamadas bases.

**probability** The chance that some event will happen. It is the ratio of the number of ways a certain event can occur to the number of possible outcomes.

**probabilidad** La posibilidad de que suceda un evento. Es la razón del número de maneras en que puede ocurrir un evento al número total de resultados posibles.

**proof** A logical argument in which each statement that is made is supported by a statement that is accepted as true.

**prueba** Argumento lógico en el cual cada enunciado hecho se respalda con un enunciado que se acepta como verdadero.

**property** A statement that is true for any numbers.

**propiedad** Enunciado que se cumple para cualquier número.

**pyramid** A polyhedron with one base that is a polygon and three or more triangular faces that meet at a common vertex.

**pirámide** Un poliedro con una base que es un polígono y tres o más caras triangulares que se encuentran en un vértice común.

**Pythagorean Theorem** In a right triangle, the square of the length of the hypotenuse $c$ is equal to the sum of the squares of the lengths of the legs $a$ and $b$.
$$a^2 + b^2 = c^2$$

**Teorema de Pitágoras** En un triángulo rectángulo, el cuadrado de la longitud de la hipotenusa es igual a la suma de los cuadrados de las longitudes de los catetos.
$$a^2 + b^2 = c^2$$

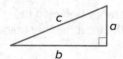

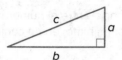

**quadrants** The four sections of the coordinate plane.

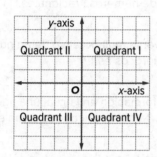

**cuadrantes** Las cuatro secciones del plano de coordenadas.

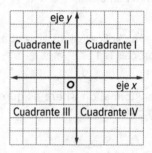

**quadratic function** A function in which the greatest power of the variable is 2.

**función cuadrática** Función en la cual la potencia mayor de la variable es 2.

**quadrilateral** A closed figure with four sides and four angles.

**cuadrilátero** Figura cerrada con cuatro lados y cuatro ángulos.

**qualitative graph** A graph used to represent situations that do not necessarily have numerical values.

**gráfica cualitativa** Gráfica que se usa para representar situaciones que no tienen valores numéricos necesariamente.

**quantitative data** Data that can be given a numerical value.

**datos cualitativos** Datos que se pueden dar un valor numérico.

**quartiles** Values that divide a set of data into four equal parts.

**cuartiles** Valores que dividen un conjunto de datos en cuatro partes iguales.

**radical sign** The symbol used to indicate a positive square root, $\sqrt{\phantom{x}}$ .

**signo radical** Símbolo que se usa para indicar una raíz cuadrada no positiva, $\sqrt{\phantom{x}}$ .

**radius** The distance from the center of a circle to any point on the circle.

**radio** Distancia desde el centro de un círculo hasta cualquier punto del mismo.

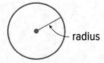

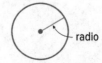

**random** Outcomes occur at random if each outcome is equally likely to occur.

**azar** Los resultados ocurren al azar si todos los resultados son equiprobables.

**range** The set of *y*-coordinates in a relation.

**rango** Conjunto de coordenadas y en una relación.

**range** The difference between the greatest number (maximum) and the least number (minimum) in a set of data.

**rango** La diferencia entre el número mayor (máximo) y el número menor (mínimo) en un conjunto de datos.

**rational number** Numbers that can be written as the ratio of two integers in which the denominator is not zero. All integers, fractions, mixed numbers, and percents are rational numbers.

**número racional** Números que pueden escribirse como la razón de dos enteros en los que el denominador no es cero. Todos los enteros, fracciones, números mixtos y porcentajes son números racionales.

**real numbers** The set of rational numbers together with the set of irrational numbers.

**número real** El conjunto de números racionales junto con el conjunto de números irracionales.

**reciprocals** The multiplicative inverse of a number. The product of reciprocals is 1.

**recíproco** El inverso multiplicativo de un número. El producto de recíprocos es 1.

**reflection** A transformation where a figure is flipped over a line. Also called a flip.

**reflexión** Transformación en la cual una figura se voltea sobre una recta. También se conoce como simetría de espejo.

**regular polygon** A polygon that is equilateral and equiangular.

**polígono regular** Polígono equilátero y equiangular.

**regular pyramid** A pyramid whose base is a regular polygon.

**pirámide regular** Pirámide cuya base es un polígono regular.

**relation** Any set of ordered pairs.

**relación** Cualquier conjunto de pares ordenados.

**relative frequency** The ratio of the number of experimental successes to the total number of experimental attempts.

**frecuencia relativa** Razón del número de éxitos experimentales al número total de intentos experimentales.

**remote interior angles** The angles of a triangle that are not adjacent to a given exterior angle.

**ángulos internos no adyacentes** Ángulos de un triángulo que no son adya centes a un ángulo exterior dado.

**repeating decimal** Decimal form of a rational number.

**decimal periódico** Forma decimal de un número racional.

**rhombus** A parallelogram with four congruent sides.

**rombo** Paralelogramo con cuatro lados congruentes.

**right angle**  An angle whose measure is exactly 90°.

**ángulo recto**  Ángulo que mide exactamente 90°.

**right triangle**  A triangle with one right angle.

**triángulo rectángulo**  Triángulo con un ángulo recto.

**rise**  The vertical change between any two points on a line.

**elevación**  El cambio vertical entre cualquier par de puntos en una recta.

**rotation**  A transformation in which a figure is turned about a fixed point.

**rotación**  Transformación en la cual una figura se gira alrededor de un punto fijo.

**rotational symmetry**  A type of symmetry a figure has if it can be rotated less than 360° about its center and still look like the original.

**simetría rotacional**  Tipo de simetría que tiene una figura si se puede girar menos que 360° en torno al centro y aún sigue viéndose como la figura original.

**run**  The horizontal change between any two points on a line.

**carrera**  El cambio horizontal entre cualquier par de puntos en una recta.

---

**Ss**

**sales tax**  An additional amount of money charged on certain goods and services.

**impuesto sobre las ventas**  Cantidad de dinero adicional que se cobra por ciertos artículos y servicios.

**sample**  A randomly-selected group chosen for the purpose of collecting data.

**muestra**  Subconjunto de una población que se usa con el propósito de recoger datos.

**sample space**  The set of all possible outcomes of a probability experiment.

**espacio muestral**  Conjunto de todos los resultados posibles de un experimento de probabilidad.

**scale factor**  The ratio of the lengths of two corresponding sides of two similar polygons.

**factor de escala**  La razón de las longitudes de dos lados correspondientes de dos polígonos semejantes.

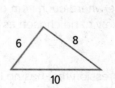

scale factor = $\frac{3}{2}$

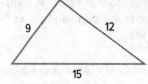

factor de escala = $\frac{3}{2}$

**scalene triangle**  A triangle with no congruent sides.

**triángulo escaleno**  Triángulo sin lados congruentes.

**scatter plot** A graph that shows the relationship between a data set with two variables graphed as ordered pairs on a coordinate plane.

**Studying for Tests**

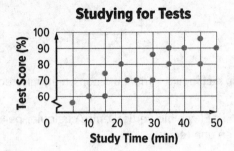

**diagrama de dispersión** Gráfica que muestra la relación entre un conjunto de datos con dos variables graficadas como pares ordenados en un plano de coordenadas.

**Tiempo de estudio para prueb**

**scientific notation** A compact way of writing numbers with absolute values that are very large or very small. In scientific notation, 5,500 is $5.5 \times 10^3$.

**notación científica** Manera abreviada de escribir números con valores absolutos que son muy grandes o muy pequeños. En notación científica, 5,500 es $5.5 \times 10^3$.

**selling price** The amount the customer pays for an item.

**precio de venta** Cantidad de dinero que paga un consumidor por un artículo.

**semicircle** An arc measuring 180°.

**semicírculo** Arco que mide 180°.

**sequence** An ordered list of numbers, such as 0, 1, 2, 3 or 2, 4, 6, 8.

**sucesión** Lista ordenada de números, tales como 0, 1, 2, 3 o 2, 4, 6, 8.

**similar** If one image can be obtained from another by a sequence of transformations and dilations.

**similar** Si una imagen puede obtenerse de otra mediante una secuencia de transformaciones y dilataciones.

**similar polygons** Polygons that have the same shape.

**polígonos semejantes** Polígonos con la misma forma.

**similar solids** Solids that have exactly the same shape, but not necessarily the same size.

**sólidos semejantes** Sólidos que tienen exactamente la misma forma, pero no necesariamente el mismo tamaño.

**simple interest** Interest paid only on the initial principal of a savings account or loan.

**interés simple** Interés que se paga sólo sobre el capital inicial de una cuenta de ahorros o préstamo.

**simple random sample** A sample where each item or person in the population is as likely to be chosen as any other.

**muestra aleatoria simple** Muestra de una población que tiene la misma probabilidad de escogerse que cualquier otra.

**simplest form** An algebraic expression that has no like terms and no parentheses.

**forma reducida** Expresión algebraica que carece de términos semejantes y de paréntesis.

**simplify** To perform all possible operations in an expression.

**simplificar** Realizar todas las operaciones posibles en una expresión.

**simulation** An experiment that is designed to model the action in a given situation.

**simulacro** Un experimento diseñado para modelar la acción en una situación dada.

**slant height** The altitude or height of each lateral face of a pyramid.

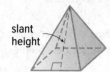

slant
height

**altura oblicua** La longitud de la altura de cada cara lateral de una pirámide.

altura
oblicua

**slope** The rate of change between any two points on a line. The ratio of the rise, or vertical change, to the run, or horizontal change.

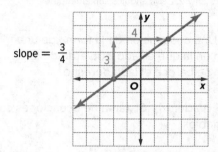

slope $= \frac{3}{4}$

**pendiente** Razón de cambio entre cualquier par de puntos en una recta. La razón de la altura, o cambio vertical, a la carrera, o cambio horizontal.

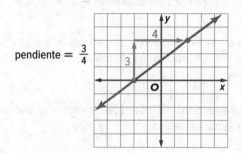

pendiente $= \frac{3}{4}$

**slope-intercept form** An equation written in the form $y = mx + b$, where $m$ is the slope and $b$ is the $y$-intercept.

**forma pendiente intersección** Ecuación de la forma $y = mx + b$, donde $m$ es la pendiente y $b$ es la intersección $y$.

**solid** A three-dimensional figure formed by intersecting planes.

**sólido** Figura tridimensional formada por planos que se intersecan.

**sphere** The set of all points in space that are a given distance from a given point called the center.

**esfera** Conjunto de todos los puntos en el espacio que están a una distancia dada de un punto dado llamado centro.

**square root** One of the two equal factors of a number. If $a^2 = b$, then $a$ is the square root of $b$. A square root of 144 is 12 since $12^2 = 144$.

**raíz cuadrada** Uno de dos factores iguales de un número. Si $a^2 = b$, la $a$ es la raíz cuadrada de $b$. Una raíz cuadrada de 144 es 12 porque $12^2 = 144$.

**standard deviation** A measure of variation that describes how the data deviates from the mean of the data.

**desviación estándar** Una medida de variación que describe cómo los datos se desvía de la media de los datos.

**standard form** An equation written in the form $Ax + By = C$.

**forma estándar** Ecuación escrita en la forma $Ax + By = C$.

**straight angle** An angle whose measure is exactly 180°.

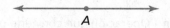

A

**ángulo llano** Ángulo que mide exactamente 180°.

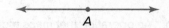

A

**substitution** An algebraic model that can be used to find the exact solution of a system of equations.

**sustitución** Modelo algebraico que se puede usar para calcular la solución exacta de un sistema de ecuaciones.

**Subtraction Property of Equality**   If you subtract the same number from each side of an equation, the two sides remain equal.

**propiedad de sustracción de la igualdad**   Si sustraes el mismo número de ambos lados de una ecuación, los dos lados permanecen iguales.

**supplementary angles**   Two angles are supplementary if the sum of their measures is 180°.

**ángulos suplementarios**   Dos ángulos son suplementarios si la suma de sus medidas es 180°.

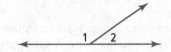

∠1 and ∠2 are supplementary angles.

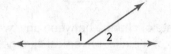

∠1 y ∠2 son ángulos suplementarios.

**symmetric**   A description of the shape of a distribution in which the left side of the distribution looks like the right side.

**simétrico**   Una descripción de la forma de una distribución en la que el lado izquierdo de la distribución se parece el lado derecho.

**system of equations**   A set of two or more equations with the same variables.

**sistema de ecuaciones**   Sistema de ecuaciones con las mismas variables.

**term**   A number, a variable, or a product of numbers and variables.

**término**   Un número, una variable o un producto de números y variables.

**term**   Each part of an algebraic expression separated by an addition or subtraction sign.

**término**   Cada parte de un expresión algebraica separada por un signo adición o un signo sustracción.

**terminating decimal**   A repeating decimal where the repeating digit is zero.

**decimal finito**   Un decimal periódico donde el dígito que se repite es cero.

**theorem**   A statement or conjecture that can be proven.

**teorema**   Un enunciado o conjetura que puede probarse.

**theoretical probability**   Probability based on known characteristics or facts.

**probabilidad teórica**   Probabilidad que se basa en características o hechos conocidos.

**third quartile**   For a data set with median M, the third quartile is the median of the data values greater than M.

**tercer cuartil**   Para un conjunto de datos con la mediana M, el tercer cuartil es la mediana de los valores mayores que M.

**three-dimensional figure**   A figure with length, width, and height.

**figura tridimensional**   Figura que tiene largo, ancho y alto.

**total surface area**   The sum of the areas of the surfaces of a solid.

**área de superficie total**   La suma del área de las superficies de un sólido.

**transformation**   An operation that maps a geometric figure, preimage, onto a new figure, image.

**transformación**   Operación que convierte una figura geométrica, la pre-imagen, en una figura nueva, la imagen.

**translation**  A transformation that slides a figure from one position to another without turning.

**traslación**  Transformación en la cual una figura se desliza de una posición a otra sin hacerla girar.

**transversal**  A line that intersects two or more other lines.

**transversal**  Recta que interseca dos o más rectas.

**trapezoid**  A quadrilateral with exactly one pair of parallel sides.

**trapecio**  Cuadrilátero con exactamente un par de lados paralelos.

**tree diagram**  A diagram used to show the total number of possible outcomes in a probability experiment.

**diagrama de árbol**  Diagrama que se usa para mostrar el número total de resultados posibles en un experimento de probabilidad.

**triangle**  A figure formed by three line segments that intersect only at their endpoints.

**triángulo**  Figura formada por tres segmentos de recta que se intersecan sólo en sus extremos.

**two-column proof**  A formal proof that contains statements and reasons organized in two columns. Each step is called a statement, and the properties that justify each step are called reasons.

**demostración de dos columnas**  Demostración formal que contiene enunciados y razones organizadas en dos columnas. Cada paso se llama enunciado y las propiedades que lo justifican son las razones.

**two-step equation**  An equation that contains two operations.

**ecuación de dos pasos**  Ecuación que contiene dos operaciones.

**two-step inequality**  An inequality that contains two operations.

**desigualdad de dos pasos**  Desigualdad que contiene dos operaciones.

**two-way table**  A table that shows data that pertain to two different categories.

**tabla de doble entrada**  Una tabla que muestra datos que pertenecen a dos categorías diferentes.

---

**Uu**

**unbiased sample**  A sample that is selected so that it is representative of the entire population.

**muestra no sesgada**  Muestra que se selecciona de modo que sea representativa de la población entera.

**unit rate/ratio**  A rate or ratio with a denominator of 1.

**tasa/razón unitaria**  Una tasa o razón con un denominador de 1.

**univariate data**  Data with one variable.

**datos univariante**  Datos con una variable.

**unlike fractions**  Fractions whose denominators are different.

**fracciones con distinto denominador**  Fracciones cuyos denominadores son diferentes.

---

**Vv**

**variable**  A symbol, usually a letter, used to represent a number in mathematical expressions or sentences.

**variable**  Un símbolo, por lo general, una letra, que se usa para representar números en expresiones o enunciados matemáticos.

**vertex**   The point where the sides of an angle meet.

**vértice**   Punto donde se encuentran los lados.

**vertex**   The point where three or more faces of a polyhedron intersect.

**vértice**   El punto donde tres o más caras de un poliedro se cruzan.

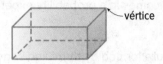

**vertex**   The point at the tip of a cone.

**vértice**   El punto en la punta de un cono.

**vertical angles**   Opposite angles formed by the intersection of two lines. Vertical angles are congruent. In the figure, the vertical angles are ∠1 and ∠3, and ∠2 and ∠4.

**ángulos opuestos por el vértice**   Ángulos congruentes que se forman de la intersección de dos rectas. En la figura, los ángulos opuestos por el vértice son ∠1 y ∠3, y ∠2 y ∠4.

**volume**   The measure of the space occupied by a solid. Standard measures are cubic units such as in³ or ft³.

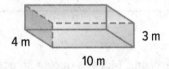

$V = 10 \times 4 \times 3 = 120$ cubic meters

**volumen**   Medida del espacio que ocupa un sólido. Las medidas estándares son las unidades cúbicas, como pulg³ o pies³.

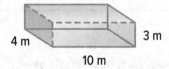

$V = 10 \times 4 \times 3 = 120$ metros cúbicos

**voluntary response sample**   A sample which involves only those who want to participate in the sampling.

**muestra de respuesta voluntaria**   Muestra que involucra sólo aquellos que quieren participar en el muestreo.

**x-axis**   The horizontal number line that helps to form the coordinate plane.

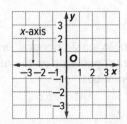

**eje x**   La recta numérica horizontal que ayuda a formar el plano de coordenadas.

**x-coordinate**   The first number of an ordered pair.

**x-intercept**   The x-coordinate of the point where the line crosses the x-axis.

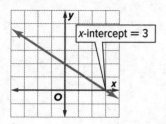

**coordenada x**   El primer número de un par ordenado.

**intersección x**   La coordenada x del punto donde cruza la gráfica el eje x.

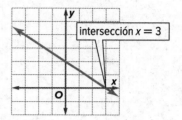

---

**Yy**

**y-axis**   The vertical number line that helps to form the coordinate plane.

**eje y**   La recta numérica vertical que ayuda a formar el plano de coordenadas.

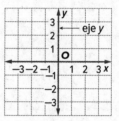

**y-coordinate**   The second number of an ordered pair.

**y-intercept**   The y-coordinate of the point where the line crosses the y-axis.

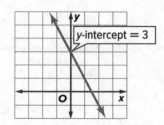

**coordenada y**   El segundo número de un par ordenado.

**intersección y**   La coordenada y del punto donde cruza la gráfica el eje y.

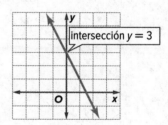

## Chapter 5 Triangles and the Pythagorean Theorem

*Page 368  Chapter 5  Are You Ready?*

**1.** 86  **3.** 98

**5–9.**

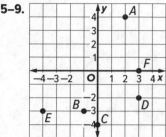

*Pages 375–376  Lesson 5-1  Independent Practice*

**1.** corresponding  **3** $m\angle 4 = 30°$, $m\angle 7 = 150°$; Sample answer: $\angle 1$ and $\angle 7$ are corresponding angles so their measures are equal. $\angle 1$ and $\angle 4$ are supplementary. So, $m\angle 4 = 180° - 150°$ or 30°.  **5.** 110°; Sample answer: $\angle 2$ and $\angle 8$ are alternate interior angles, so they have the same measure.  **7a.** 20  **7b.** 40  **9 a.** The top and bottom of the ramp are parallel. The slanted part of the ramp can be considered a transversal. You can use angle relationships of parallel lines to find the measure of the missing angle.
**b.** 28°  **11.** They are supplementary.

*Pages 377–378  Lesson 5-1  Extra Practice*

**13.** alternate interior  **15.** alternate interior  **17.** 110°; Sample answer: $\angle 2$ and $\angle 6$ are corresponding angles, so they have the same measure.  **19.** 43°; Sample answer: $\angle 11$ and $\angle 3$ are corrresponding angles, so they have the same measure and $\angle 3$ and $\angle 4$ are supplementary. So, $m\angle 4 = 180 - 137$ or 43°.  **21.** $\angle 2$ and $\angle 6$; $\angle 1$ and $\angle 5$  **23.** 16 in$^2$
**25.** supplementary

*Pages 383–384  Lesson 5-2  Independent Practice*

**1** Proof: $\angle 7$ and $\angle 8$ form a straight angle so they are supplementary angles. So, $m\angle 7 + m\angle 8 = 180°$, by the definition of supplementary angles. By substitution, $9x + 11x = 180$. So, $x = 9$ by the Division Property of Equality.

**3.**

| | Statements | Reasons |
|---|---|---|
| a. | $\angle 1$ and $\angle 2$ are supplementary; $m\angle 1 = m\angle 2$ | Given |
| b. | $m\angle 1 + m\angle 2 = 180°$ | Definition of supplementary angles |
| c. | $m\angle 1 + m\angle 1 = 180°$ | Substitution |
| d. | $2(m\angle 1) = 180°$ | Simplify. |
| e. | $m\angle 1 = 90°$ | Division Property of Equality |
| f. | $m\angle 2 = 90°$ | $m\angle 1 = m\angle 2$ (Given) |
| g. | $\angle 1$ and $\angle 2$ are right angles. | Definition of right angles |

**5.** Sample answer: Vertical angles have the same measure.
**7.** ll

*Pages 385–386  Lesson 5-2  Extra Practice*

**9.**

| | Statements | Reasons |
|---|---|---|
| a. | $j \parallel k$, transversal $\ell$; $m\angle 3 = 2x - 15$, $m\angle 6 = x + 55$ | Given |
| b. | $m\angle 3 = m\angle 6$ | Alternate interior angles have the same measure. |
| c. | $2x - 15 = x + 55$ | Substitution |
| d. | $x - 15 = 55$ | Subtraction Property of Equality |
| e. | $x = 70$ | Addition Property of Equality |

**11a.** Yes  **11b.** Yes  **11c.** Yes  **11d.** No  **13.** adjacent
**15.** neither

*Pages 393–394  Lesson 5-3  Independent Practice*

**1.** 55  **3.** 24°, 48°, 108°  **5.** 112  **7.** 45  **9** 105°
**11** 90°, 60°, 30°  **13.** Sample answer: Since $\angle 1$ and $\angle 4$ form a straight angle, $m\angle 1 + m\angle 4 = 180°$. By the Subtraction Property of Equality, $m\angle 1 = 180 - m\angle 4$. Since $ABC$ is a triangle, $m\angle 2 + m\angle 3 + m\angle 4 = 180$. By the Subtraction Property of Equality, $m\angle 2 + m\angle 3 = 180 - m\angle 4$. So by substitution, $m\angle 2 + m\angle 3 = m\angle 1$.  **15.** Sample answer: The sum is 360°. Drawing the diagonal of a quadrilateral forms two triangles. So, the sum of the interior angles is 2(180°), or 360°.

**17.** 100  **19.** 50  **21.** 120  **23.** 48°, 60°, 72°  **25.** 70
**27.** 25; 50  **29.** They are acute; They are complementary.
**31.** 51°  **33.** 41

**1.** 540°  **3** 1,980°  **5** 36°  **7.** 24°  **9.** 60°, 90°, 90°, 120°;
360°  **11.** 130
**13.** 18

$$\frac{(n-2)180}{n} = 160$$
$$(n-2)180 = 160n \quad \text{Multiplication Property}$$
$$\text{of Equality}$$
$$180n - 360 = 160n \quad \text{Distributive Property}$$
$$20n = 360 \quad \text{Properties of Equality}$$
$$n = 18 \quad \text{Division Property of}$$
$$\text{Equality}$$

**15.** Regular decagons have equal angles measuring 144° and
regular 11-sided polygons have angles measuring 147.27°.
145° is between these two values so it cannot be the interior
angle measure of a regular polygon.

**17.** 2,160°  **19.** 140°  **21.** 161.1°  **23.** 40°  **25.** 20°
**27.** Sample answer: The sum of the interior angles will still be
720° because even though the figures are not regular, they
are still hexagons.  **29a.** 360°  **29b.** hexagon  **29c.** 1,080°
**29d.** 11-gon  **31.** neither  **33.** 26

**Case 3.** 1,800°  **Case 5.** 1,234,567,654,321

**1.** $5^2 + 12^2 = c^2$; 13 in.  **3.** $8^2 + b^2 = 18^2$; 16.1 m
**5** no; $30^2 + 122^2 \neq 125^2$  **7** $48^2 + 55^2 = c^2$; 73 yd
**9.** $a^2 + 5.1^2 = 12.3^2$; 11.2 m  **11.** 7.8 cm  **13.** about 7.1 in;
Sample answer: The Pythagorean Theorem states that
$c^2 = a^2 + b^2$. Since both legs are $x$ inches, $c^2 = 2x^2$. When
you replace $c$ with 10 and simplify, $x \approx 7.1$.

**15.** $27^2 + 36^2 = c^2$; 45 ft  **17.** $30^2 + b^2 = 80^2$; 74.2 mm
**19.** no; $135^2 + 140^2 \neq 175^2$  **21.** no; $44^2 + 55^2 \neq 70^2$
**23.** Sample answer: If you know the lengths of two sides of a
right triangle, you can substitute the values in the Pythagorean
Theorem and find the missing side.  **25.** 8 cm, 15 cm, 17 cm;
6 cm, 8 cm, 10 cm  **27.** 260  **29.** 64 units  **31.** 6.7

**1** $5^2 + h^2 = 12^2$; 10.9 ft  **3** 11.7 cm  **5a.** 40 yd
**5b.** 24.7 yd  **9.** 3–5–7; $3^2 + 5^2 \neq 7^2$

**13.** 9.0 in.  **15.** about 105 mi  **17.** 20.6 in.  **19.** 2.5 in.
**21.** 6; $6^2 = 36$ and $7^2 = 49$, since 39 is closer to 36 than 49,
$\sqrt{39} \approx 6$.  **23.** 3; $3^3 = 27$ and $4^3 = 64$. Since 30 is closer to
27 than 64, $\sqrt[3]{30} \approx 3$.

**1.**
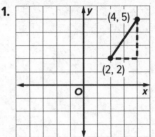
3.6 units

**3**
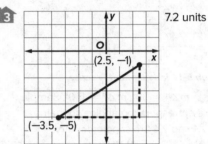
7.2 units

**5** 1.4 units  **7.** 15.9 units
**9a.**
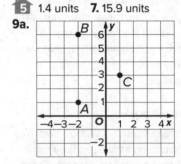

**9b.** Sample answer: Use the Distance Formula and the points
(−2, 6) and (1, 3).  **9c.** 3.6 units; 5 units; 4.2 units  **9d.** 12.8 units
**11.** Sample answer: (1, 2) and (4, 6)

**13.**
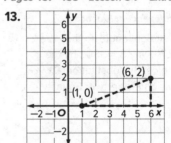
5.4 units

**15.** 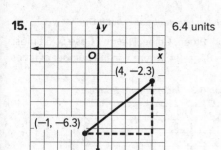 6.4 units

(4, −2.3)

(−1, −6.3)

**17.** 11.4 units  **19.** 13.6 units  **21a.** True  **21b.** True
**21c.** False

**23.** 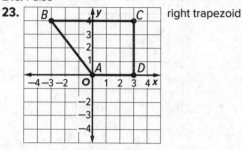 right trapezoid

*Page 441  Chapter Review  Vocabulary Check*

**1.** transversal  **3.** Deductive reasoning  **5.** Pythagorean
Theorem  **7.** parallel lines  **9.** hypotenuse

*Page 442  Chapter Review  Key Concept Check*

**1.** corresponding  **3.** 4

# Chapter 6 Transformations

*Page 448  Chapter 6  Are You Ready?*

**1.** Sample answer:

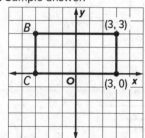

**3.**

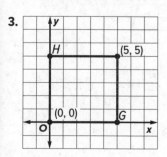

**5.** −2  **7.** 6  **9.** −6  **11.** 2

*Pages 457–458  Lesson 6-1  Independent Practice*

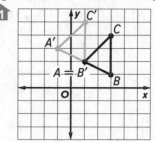

$A'(-1, 3)$, $B'(1, 2)$, $C'(1, 5)$
**3.** $P'(6, 5)$, $Q'(11, 3)$, $R'(3, 11)$  **5.** $(x - 3, y - 3)$  **7** $K''(-3, 1)$,
$L''(0, 4)$, $M''(-1, 7)$, $N''(-4, 8)$  **9.** the same as the original
position of the figure; Sample answer: Since −5 and 5 are
opposites, and −7 and 7 are opposites, the translations cancel
each other out.  **11a.** always; Sample answer: Each point
moves the same distance and in the same direction.
**11b.** never; Sample answer: A preimage and image in a
translation will always have the same size and shape.

*Pages 459–460  Lesson 6-1  Extra Practice*

**13.**

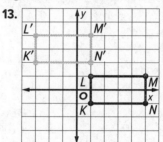

$K'(-3, 2)$, $L'(-3, 4)$, $M'(1, 4)$, $N'(1, 2)$
**15.** $A'(-3, -3)$, $B'(-1, -2)$, $C'(4, -4)$, $D'(2, -8)$
**17.**

**19.** $A'(-2, 4)$; $D'(0, 6)$  **21.** 20  **23.** 174  **25.** −355

*Pages 465–466  Lesson 6-2  Independent Practice*

**1.**

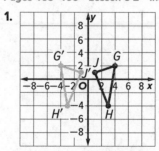

$G'(-4, 2)$, $H'(-3, -4)$, $J'(-1, 1)$
**3.**

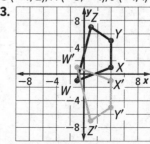

$W'(-1, 1)$, $X'(4, -1)$, $Y'(4, -5)$, $Z'(1, -7)$

**5.** $A'(-3, -3)$, $B'(3, -3)$

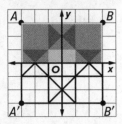

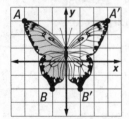 *x*-axis   **9.** $J''(7, -4)$, $K''(-7, -1)$, $L''(-2, 2)$   **11.** no; Sample answer: If the vertices of $\triangle ABC$ are $A(0, 0)$, $B(2, 2)$, and $C(0, 4)$, then the vertices of the final image are $A''(0, 0)$, $B''(-2, -2)$, and $C''(0, -4)$.

*Pages 467–468   Lesson 6-2   Extra Practice*

**13.**

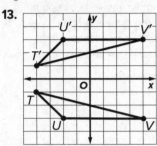

$T'(-4, 1)$, $U'(-2, 3)$, $V'(4, 3)$

**15.**

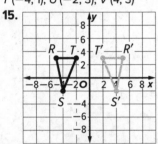

$R'(5, 3)$, $S'(4, -2)$, $T'(2, 3)$
**17.** $A'(3, 3)$, $B'(1, -2)$

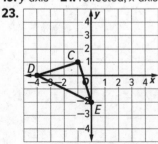

**19.** *y*-axis   **21.** reflected; *x*-axis
**23.**

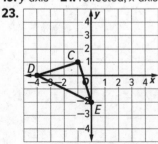

*Page 471   Problem-Solving Investigation   Act It Out*

**Case 3.** Abril, Brandon, Ethan, Chloe, Diego   **Case 5.** 6 times; Sample answer: $(20 - 8)x = 84$, where $x =$ the number of times

*Pages 479–480   Lesson 6-3   Independent Practice*

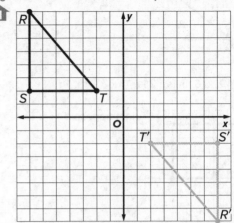

$R'(7, -8)$, $S'(7, -2)$, $T'(2, -2)$
**3.**

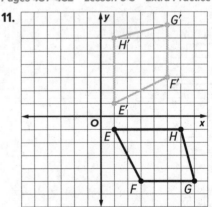

I and N   **7.** $A''(0, 4)$, $B''(0, -2)$, $C''(-2, 0)$   **9a.** $(x, y) \rightarrow (y, -x) \rightarrow (x, y)$   **9b.** They are the same as a rotation of 360°.

*Pages 481–482   Lesson 6-3   Extra Practice*

**11.**

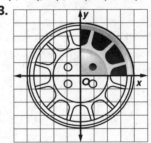

$E'(1, 1)$, $F'(5, 3)$, $G'(5, 7)$, $H'(1, 6)$   **13a.** reflection
**13b.** translation   **13c.** rotation   **15.** $M'(-1, -4)$, $N'(-3, -1)$, $P'(-5, -3)$   **17.** $T'(2, 4)$, $U'(2, 2)$, $V'(-1, 2)$, $W'(-1, 4)$

**19a.** $A'(-2, 2)$, $B'(-1, -2)$, $C'(1, 0)$

**19b.**

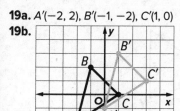

*Pages 491–492   Lesson 6-4   Independent Practice*

**1** $C'(2, 8)$, $A'(4, 4)$, $T'(10, 10)$

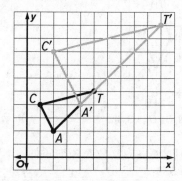

**3.** $\frac{1}{5}$   **5** **a.** $A''(-6, -9)$, $B''(0, 0)$, $C''(3, -3)$   **b.** $A''(-6, -9)$, $B''(0, 0)$, $C''(3, -3)$   **c.** Yes; Sample answer: since the coordinates of the answers to Exercises a and b are the same, the order in which you perform them does not matter. **7a.** $(-12, -18)$   **7b.** The final coordinates are three times the original coordinates.   **7c.** Sample answer: Yes; multiply the scale factors of each dilation to find the scale factor of the final dilation.   **9.** Sample answer: $a = \frac{1}{3}$, $a = \frac{1}{5}$, $a = \frac{1}{2}$

*Pages 493–494   Lesson 6-4   Extra Practice*

**11.** $V'(-9, 12)$, $X'(-6, 0)$, $W'(3, 6)$

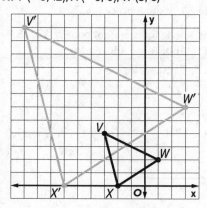

**13a.** $X'(0, 0)$, $Y'(-6, -2)$, $Z'(-4, -6)$

**13b.**

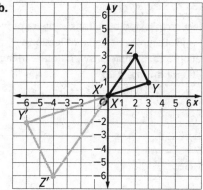

**13c.** $\triangle X'Y'Z'$ is the image of $\triangle XYZ$ after a dilation of 2 and a rotation of 180° about the origin.   **15a.** True   **15b.** True   **15c.** False   **17.** $\frac{1}{20}$   **19.** $\frac{1}{9}$

*Page 497   Chapter Review   Vocabulary Check*

**1.** Translation   **3.** transformation; preimage; image

*Page 498   Chapter Review   Key Concept Check*

**1.** correct   **3.** correct

# Chapter 7  Congruence and Similarity

*Page 504   Chapter 7   Are You Ready?*

**1.** 3.5   **3.** 14.8   **5.** 2   **7.** $-3$   **9.** $\frac{5}{3}$

*Pages 513–514   Lesson 7-1   Independent Practice*

**1** not congruent; No sequence of transformations maps *RSTU* onto *WXYZ* exactly.   **3** Sample answer: 90° clockwise rotation followed by a translation; they are congruent because an image produced by a rotation and a translation have the same size and shape.

**5.**

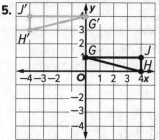

4 units, 1 unit, $\sqrt{17}$ units; 4 units, 1 unit, $\sqrt{17}$ units; yes   **7.** $A(-3, 4)$, $B(-2, 1)$, $C(2, 2)$   **9a.** true; Sample answer: The segment was translated x units to the right.   **9b.** false; Sample answer: The segment was dilated by a scale factor of $\frac{2}{3}$.

**11.** congruent; A reflection followed by a translation maps △*FGH* onto △*MNP*.    **13.** Sample answer: a reflection followed by a translation    **15.** Sample answer: Rotate triangle *A* counterclockwise, then translate it up.

**17.** 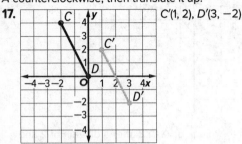 *C'*(1, 2), *D'*(3, −2)

**1** ∠*N* ≅ ∠*S*, ∠*M* ≅ ∠*T*, ∠*O* ≅ ∠*V*; $\overline{ON} ≅ \overline{VS}$, $\overline{NM} ≅ \overline{ST}$, $\overline{MO} ≅ \overline{TV}$    **3** ∠*U* ≅ ∠*H*, ∠*V* ≅ ∠*J*, ∠*W* ≅ ∠*I*, ∠*X* ≅ ∠*K*; $\overline{UV} ≅ \overline{HJ}$, $\overline{VW} ≅ \overline{JI}$, $\overline{WX} ≅ \overline{IK}$, $\overline{XU} ≅ \overline{KH}$ ; Sample answer: If you reflect parallelogram *UVWX* over the *x*-axis, then translate it 4 units to the right, it coincides with parallelogram *HJIK*.

**5a.**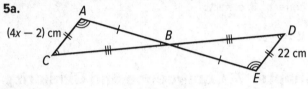

**5b.** 6

**7a.**

**7b.** 4    **7c.** △*ABC* ≅ △*CDE* and △*CAF* ≅ △*CEF*.
**9a.** true; Sample answer: If the figures are congruent, the corresponding sides have equal length. Therefore, the sum of the lengths of the sides will be equal.    **9b.** false; Sample answer: Triangle *ABC* has a perimeter of 24 inches. Square *MNOP* has a perimeter of 24 inches. They have the same perimeter but because they are different shapes, they are not congruent.

**11.** ∠*S* ≅ ∠*Y*, ∠*STZ* ≅ ∠*YTW*, ∠*Z* ≅ ∠*W*; $\overline{SZ} ≅ \overline{YW}$, $\overline{ZT} ≅ \overline{WT}$, $\overline{TS} ≅ \overline{TY}$    **13.** ∠*K* ≅ ∠*F*, ∠*L* ≅ ∠*G*, ∠*M* ≅ ∠*H*, ∠*N* ≅ ∠*J*; $\overline{KL} ≅ \overline{FG}$, $LM ≅ \overline{GH}$, $\overline{MN} ≅ \overline{HJ}$, $\overline{NK} ≅ \overline{JF}$; Sample answer: If you reflect quadrilateral *KLMN* over the *y*-axis, then translate it 5 units up and 2 units to the left, it coincides with quadrilateral *FGHJ*.

**15a.**

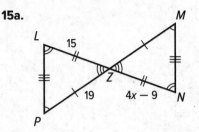

**15b.** 6    **17.** $\overline{RP} ≅ \overline{RS}$; ∠*SRQ* ≅ ∠*PRQ*
**19.**

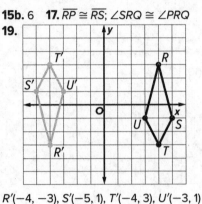

*R'*(−4, −3), *S'*(−5, 1), *T'*(−4, 3), *U'*(−3, 1)

**Case 3.** 20 people    **Case 5.** 3

**1** yes; Sample answer: A rotation, a translation of 4 units down, and a dilation with a scale factor of $\frac{3}{2}$ maps △*XYW* onto △*VUW*.    **3** 6.75 in. by 11.25 in.; yes    **5.** Sample answer: translation of 1 unit to the right and 1 unit down followed by a dilation with a scale factor of 4    **7.** Product of dilation(s) should equal 1.    **9.** false; Sample answer: If you perform the dilation after a translation, the translation is multiplied by the same scale factor.

**11.** no; The ratios of the side lengths are not equal.    **13.** 7.5 ft by 6 ft; yes    **15a.** False    **15b.** True    **15c.** False

**17.**

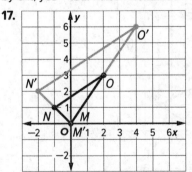

*M'*(0, 0), *N'*(−2, 2), *O'*(4, 6)

**19.**

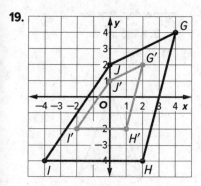

$G'(2, 2)$, $H'(1, -2)$, $I'(-1, -2)$, $J'(0, 1)$

*Pages 549–550   Lesson 7-4   Independent Practice*

**1** No; The corresponding angles are congruent, but $\frac{3}{7} \neq \frac{4}{8}$.   **3** translation and dilation; 4.5   **5a.** Figure 1: 96 cm$^2$; Figure 2: 294 cm$^2$   **5b.** Sample answer: The scale factor of the side lengths is $\frac{14}{8}$ or $\frac{7}{4}$. The ratio of the areas is $\frac{49}{16}$. The ratio of the areas is the scale factor of the side lengths squared.   **7.** 400 ft   **9.** false; Sample answer: In rectangles, all corresponding angles are congruent but not all sides are proportional. Rectangle A is not similar to Rectangle B, since $\frac{4}{4} \neq \frac{1}{2}$.

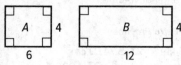

*Pages 551–552   Lesson 7-4   Extra Practice*

**13.** No; the corresponding angles are congruent, but $\frac{5}{4} \neq \frac{8}{6}$.   **15.** 70 ft   **17a.** True   **17b.** True   **17c.** False   **19.** $\frac{1}{24}$   **21.** $\frac{1}{2}$   **23.** $\frac{1}{7,920}$

*Pages 557–558   Lesson 7-5   Independent Practice*

**1.** The triangles are not similar.   **3.** 200 ft   **5** 37.5 m   **7** $\frac{136}{34} = \frac{h}{1.5}$; 6 feet tall   **11.** Sample answer: The length of the tall object's shadow, the length of the shadow of a nearby object with a height that is directly measurable, and the height of the nearby object.

*Pages 559–560   Lesson 7-5   Extra Practice*

**13.** 90 ft   **15.** 6 m   **17a.** $\frac{h}{ED} = \frac{BC}{DC}$   **17b.** The distance from the mirror to the person, the distance from the mirror to the base of the flag, the height of the person's eyes.   **19.** Sample answer: $\frac{6\,\text{ft}}{3\,\text{ft}} = \frac{h\,\text{ft}}{25\,\text{ft}}$; 50   **21.** Yes; the corresponding angles are congruent and $\frac{5}{10} = \frac{4}{8}$.

*Pages 565–566   Lesson 7-6   Independent Practice*

**1** 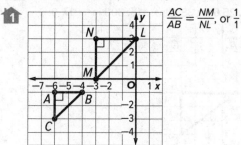 $\frac{AC}{AB} = \frac{NM}{NL}$, or $\frac{1}{1}$

**3** $m = -\frac{2}{5}$; The other slope should equal $-\frac{2}{5}$.   **5.** $P(5, 3)$

**9a.** similar triangle, slope triangle   **9b.** similar triangle   **9c.** neither   **9d.** similar triangle, slope triangle

*Pages 567–568   Lesson 7-6   Extra Practice*

**11.** 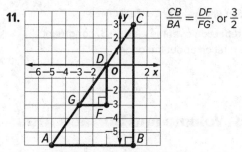 $\frac{CB}{BA} = \frac{DF}{FG}$, or $\frac{3}{2}$

**13.** 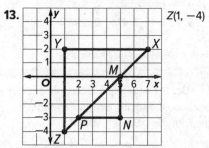 $Z(1, -4)$

**15.** $D(1, 11)$   **17a.** True   **17b.** True   **17c.** True   **19.** 0   **21.** undefined   **23.** $\frac{5}{3}$   **25.** $\frac{1}{3}$

*Pages 573–574   Lesson 7-7   Independent Practice*

**1** 57 mm   **3.** 160 ft   **5** 126 ft$^2$

**7.**

| If the scale factor is... | Multiply the ... | | | |
| --- | --- | --- | --- | --- |
| | Length by | Width by | Perimeter by | Area by |
| 2 | 2 | 2 | 2 | 4 |
| 4 | 4 | 4 | 4 | 16 |
| 0.5 | 0.5 | 0.5 | 0.5 | 0.25 |
| $\frac{2}{3}$ | $\frac{2}{3}$ | $\frac{2}{3}$ | $\frac{2}{3}$ | $\frac{4}{9}$ |
| $k$ | $k$ | $k$ | $k$ | $k^2$ |

**9.** Robert is thinking of size in terms of area and Denise is thinking of size in terms of perimeter.

Pages 575–576   Lesson 7-7   *Extra Practice*

**11.** 30 cm   **13.** 25.6 in.; 38.4 in$^2$   **15.** 300 in$^2$   **17.** 9; 3

**19.**

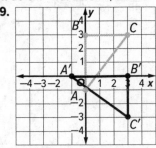

Page 579   *Chapter Review   Vocabulary Check*

**1.** congruent   **3.** corresponding parts   **5.** composition of transformations

Page 580   *Chapter Review   Key Concept Check*

**1.** congruent; a reflection over the *x*-axis   **3.** congruent; a 90° clockwise rotation about the origin

# Chapter 8   Volume and Surface Area

Page 586   *Chapter 8   Are You Ready?*

**1.** 68 cm$^2$   **3.** 71.5 m$^2$   **5.** 72 in$^2$   **7.** 47.1   **9.** 283.4

Pages 593–594   Lesson 8-1   *Independent Practice*

**1** 141.4 in$^3$   **3** 831.9 lb   **5a.** bag: 132 in$^3$; candle: 29.5 in$^3$   **5b.** 102.5 in$^3$   **5c.** 13 packages   **7.** Sample answer: The shorter cylinder, because the radius is larger and that is the squared value in the formula.   **9a.** 2:1   **9b.** 4:1

Pages 595–596   Lesson 8-1   *Extra Practice*

**11** 2,770.9 yd$^3$   **13.** 81.7 ounces   **15.** 8 in.
**17a.** $V = \pi(1)^2(1)$; $V = \pi(1)^2(2)$; $V = \pi(2)^2(1)$; $V = \pi(2)^2(2)$
**17b.** The height of Cylinder B is twice the height of Cylinder A. The radius of Cylinder C is twice the radius of Cylinder A. The radius and height of Cylinder D are twice the radius and height of Cylinder A.
**17c.**

| | Radius (cm) | Height (cm) | Volume (cm$^3$) |
|---|---|---|---|
| **Cylinder A** | 1 | 1 | 3.14 cm$^3$ |
| **Cylinder B** | 1 | 2 | 6.28 cm$^3$ |
| **Cylinder C** | 2 | 1 | 12.57 cm$^3$ |
| **Cylinder D** | 2 | 2 | 25.13 cm$^3$ |

**17d.** When the radius is doubled, the volume is four times the original volume. When the height is doubled, the volume is twice the original volume. When the radius and height are doubled, the volume is eight times the original volume.
**19.** The volume of the container is exactly 20.25$\pi$ cubic inches; The volume of the container to the nearest tenth is about 63.6 cubic inches.   **21.** 201.1 cm$^2$   **23.** 28.3 in$^2$
**25.** 50.3 m$^2$   **27.** 539 m$^3$

Pages 601–602   Lesson 8-2   *Independent Practice*

**1.** 4,720.8 mm$^3$   **3** 26.9 ft$^3$   **5.** 102.6 in$^3$
**7.** 1,608.5 cm$^3$   **9** 36 cm   **11.** 10 mm
**13.** Sample answer:

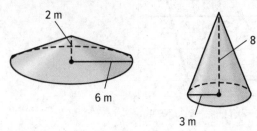

Pages 603–604   Lesson 8-2   *Extra Practice*

**15** 2,989.8 mm$^3$   **17.** 398.2 m$^3$   **19.** 402.1 cm$^3$
**21.** Sample answer: 32.7 in$^3$   **23.** 15 in.   **25.** 4.5 m
**27.** 3.0 yd   **29a.** True   **29b.** False   **29c.** True   **31.** 110 ft$^3$
**33.** 90 cm$^3$

Pages 609–610   Lesson 8-3   *Independent Practice*

**1** 1,563.5 in$^3$   **3.** 2,144.7 mm$^3$   **5** 434.9 in$^3$
**7.** 107.2 s   **9.** 1.5 mm$^3$   **11.** 10.9 in.   **13.** 91.8 cm$^3$
**15.** false; Sample answer: The radius is cubed when finding the volume of a sphere. When the radius is doubled, the volume is 2$^3$ or 8 times the original volume.

Pages 611–612   Lesson 8-3   *Extra Practice*

**17** 3,053.6 in$^3$   **19.** 883.6 km$^3$   **21.** 5,747.0 cm$^3$
**23.** 5,575.3 cm$^3$   **25.** 0.0011 grams/mm$^3$   **27.** 1,038.2 cm$^3$
**29a.** 4.5   **29b.** 381.7   **29c.** 190.9   **31.** 25.1 mm; 50.3 mm$^2$
**33.** 19.5 m; 30.2 m$^2$   **35.** 134.8 cm$^2$

Page 615   *Problem-Solving Investigation   Solve a Simpler Problem*

**Case 3.** 389.6 ft$^3$   **Case 5.** 55 squares

Pages 623–624   Lesson 8-4   *Independent Practice*

**1** 88.0 mm$^2$   **3.** 272.0 mm$^2$   **5.** 113.1 in$^2$   **7.** 1,068.1 yd$^2$
**9** 241.3 in$^2$   **11.** No, the surface area of the side of the cylinder will double, but the area of the bases will not.
**13.** $A = 2\pi rh + \pi r^2$; Sample answer: The baker will not ice the bottom of the cake, so you only need to include the area of one of the bases in the equation.

Pages 625–626   Lesson 8-4   *Extra Practice*

**15** 1,105.8 cm$^2$; 1,508.0 cm$^2$   **17.** 763.4 in$^2$
**19.** Sample answer: 2 · 3 · 4$^2$ + 2 · 3 · 4 · 4 or 192 m$^2$
**21.** about 85.7%   **23a.** 562   **23b.** 653   **23c.** II; 91; I
**25.** 23.08 ft$^2$   **27.** 3.5 in$^3$

Pages 635–636   Lesson 8-5   *Independent Practice*

**1** 269.2 in$^2$   **3.** 785.4 m$^2$   **5.** 279.5 cm$^2$   **7** 13.4 in$^2$
**9a.** 510.2 mm$^2$   **9b.** 14.2 mm   **11.** Enrique did not use the right radius. He did not divide the diameter by 2 to get the radius; 267.04 in$^2$   **13.** square pyramid; Sample answer: The surface area of the pyramid is $x^2 + 2x\ell$. If you use $\pi \approx 3.14$,

the surface area of the cone is $0.785x^2 + 1.57x\ell$. For all positive values of $x$ and $\ell$, the surface area of the pyramid is greater than the surface area of the cone.

*Pages 637–638  Lesson 8-5  Extra Practice*

**15** 461.8 m² **17.** 113.1 in² **19.** 62.8 m² **21.** 452.4 in² **23.** 188.5 yd² **25.** 354.1 ft² **27.** The slant height of the cone is 13 cm; The lateral area of the cone is about 204 square centimeters. **29.** 150.8 ft² **31.** 829.4 cm² **33.** 1,583.4 ft² **35.** 1,742.5 ft³

*Pages 645–646  Lesson 8-6  Independent Practice*

**1.** 1,520 cm² **3** 548.8 in² **5.** 19 mm³ **7.** 8,709,120 ft³; 277,632 ft² **9** 300.8 m³ **11.** The volume of the first cone is $x$, so the first cone's volume multiplied by one-sixth cubed is the second cone's volume. The volume of the second cone is $\frac{1}{216}x$ in³.

*Pages 647–648  Lesson 8-6  Extra Practice*

**13** 2,700 ft³ **15.** 10 cm³ **17.** 0.65 m² **19.** sometimes **21.** always **23a.** 3:1 **23b.** surface area, 9:1; volume, 27:1 **23c.** 602.88 cm² **23d.** 30,520.8 cm³ **25a.** True **25b.** False **25c.** False **27a.** 54,000 ft² **27b.** 13,050 ft² **27c.** about 5.6 acres **29.** 100.7 cm²

*Page 651  Chapter Review  Vocabulary Check*

**1.** sphere **3.** cylinder **5.** cone: a three-dimensional figure with one circular base

*Page 652  Chapter Review  Key Concept Check*

**1.** correct

# Chapter 9  Scatter Plots and Data Analysis

*Page 662  Chapter 9  Are You Ready?*

**1.** Sample answer: There are 8 different kinds of mammals shown. Most of them have a life span between 10 and 29 years; 4 **3.** 31.6

*Pages 671–672  Lesson 9-1  Independent Practice*

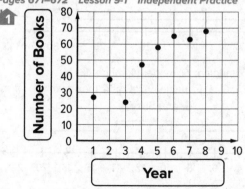

**3a.**

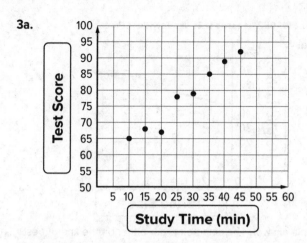

**3b.** Sample answer: The scatter plot shows a positive linear association. There are no clusters or outliers. **3c.** about 98 **5.** positive
**7.**

| Side Length (units) | Perimeter (units) | Area (units²) |
|---|---|---|
| 1 | 4 | 1 |
| 2 | 8 | 4 |
| 3 | 12 | 9 |
| 4 | 16 | 16 |
| 5 | 20 | 25 |
| 6 | 24 | 36 |

side length and perimeter; Sample answer: The data would form a straight line.

*Pages 673–674  Lesson 9-1  Extra Practice*

**9** a.

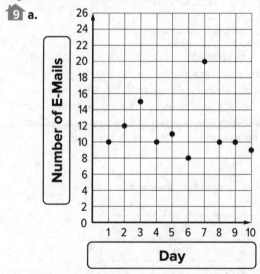

**9b.** There does not appear to be any relationship between the variables. Linearity cannot be determined and there are no clusters. There appears to be an outlier at 20 E-mails.
**9c.** Since the scatter plot shows no association between the data, it is not possible to predict how many E-mails will be received on Day 15. **11.** The time does not depend on shoe size. Therefore, the scatter plot shows no association.
**13.** increases; increases **15.** Sample answer: There are more than $2\frac{1}{2}$ times more people that speak Mandarin than English.

Pages 681–682   Lesson 9-2   Independent Practice

**1 a.**

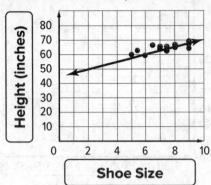

Sample answer: The data points are either on the line of best fit or very close to the line, so the line of best fit is a good model of the data.   **b.** Sample answer: 57.5 in.

**3 a.** Sample answer: $y = 500x + 2{,}250$; Every year an additional 500 girls play ice hockey. In 1996, 2,250 girls played ice hockey.   **b.** Sample answer: 14,250 girls
**7a.** always; Sample answer: A line of best fit for data with a positive association will have a positive slope.
**7b.** sometimes; Sample answer: Depending on the data, the y-intercept could be positive, negative, or zero.

Pages 683–684   Lesson 9-2   Extra Practice

**9 a.** Sample answer:

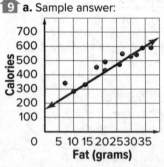

Since the data points are all close to the line, the line of best fit drawn is a good model of the data.   **9b.** Sample answer: $y = 12.5x + 155$; A 1 gram increase in fat increases the Calories by 12.5. A sandwich with 0 grams of fat would be 155 Calories.
**9c.** Sample answer: 15.6 g   **11a.** True   **11b.** True
**13.** no association

Pages 693–694   Lesson 9-3   Independent Practice

**1**

|  | Chicken | Beef | Total |
|---|---|---|---|
| Rice | 20 | 10 | 30 |
| Pasta | 40 | 30 | 70 |
| Total | 60 | 40 | 100 |

**3**

|  | Text Message | Instant Message | Total |
|---|---|---|---|
| 7th graders | 59; 0.70 | 25; 0.30 | 84; 1.00 |
| 8th graders | 59; 0.59 | 41; 0.41 | 100; 1.00 |
| Total | 118 | 66 | 184 |

Sample answer: Many more 7th grade students text message than instant message. The eighth grade is split more evenly.
**5.** Sample answer: Jasmine should have said, "More than half of the students that do not have an after-school job are on the honor roll."   **7.** the percentage of red-haired students with brown eyes

Pages 695–696   Lesson 9-3   Extra Practice

**9**

|  | Popcorn | No Popcorn | Total |
|---|---|---|---|
| Drink | 74 | 15 | 89 |
| No Drink | 10 | 6 | 16 |
| Total | 84 | 21 | 105 |

**11.** males; 26 out of 48, or 54%, of males volunteer at the animal shelter while only 21 out of 52, or 40%, of females volunteer at the animal shelter
**13.** Sample answer:

|  | Cat | No Cat | Total |
|---|---|---|---|
| Dog | 45; $\frac{45}{123} = 0.37$ | 125; $\frac{125}{177} = 0.71$ | 170 |
| No Dog | 78; $\frac{78}{123} = 0.63$ | 52; $\frac{52}{177} = 0.29$ | 130 |
| Total | 123 | 177 | 300 |

Most people that visited the store and have a cat do not have a dog. Most people that visited the store and did not have a cat, have a dog.   **15.** Of the students who attended the basketball game, fewer than half of them also attended the school play; Students who attended the school play were more likely not to attend the basketball game.   **17.** mean; Sample answer: Removing the highest score causes the mean to change from about 102 to 98. This is the greatest change.

Page 699   Problem-Solving Investigation   Use a Graph

**Case 3.** Sample answer: 208,000 followers   **Case 5.** Sample answer: 168 members

Pages 705–706   Lesson 9-4   Independent Practice

**1** mean: 103.1; median: 100; mode: 100; range: 46
**3** minimum: 20; $Q_1$: 21; median: 23.5; $Q_3$: 29; maximum: 30;

**Incubation Period**

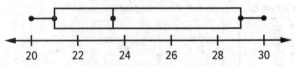

**5a.** mean: 11.6; median: 12; mode: 14; range: 16
**5b.** minimum: 3; $Q_1$: 9.5; median: 12; $Q_3$: 14; maximum: 19
**5c.**        **Menu Survey**

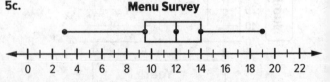

**7.** Sample answer: {1, 2, 5, 7, 9, 10, 12, 14, 15, 17, 22} and {0, 2, 5, 7, 9, 10, 12, 14, 15, 17, 27}

*Pages 707–708   Lesson 9-4   Extra Practice*

**9** mean: 72; median: 60; no mode; range: 108

**11a.** mean: 149.1; median: 154.5; mode: 162; range: 36

**11b.** minimum: 128; $Q_1$: 134; median: 154.5; $Q_3$: 162; maximum: 164

**11c.**

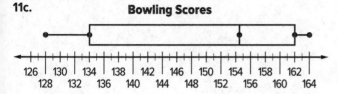

**Bowling Scores**

**13.**

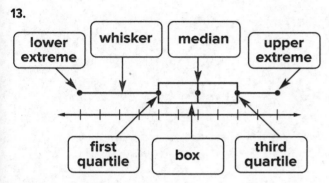

**15a.** False   **15b.** True   **15c.** False

*Pages 713–714   Lesson 9-5   Independent Practice*

**1** 8.2; Sample answer: The average distance each data value is from the mean is 8.2 miles per hour.   **3** Speeds that are between 34.2 and 56.8 miles per hour are within one standard deviation of the mean.   **5.** Sample answer: Brian should have said more than half of his data values are within one standard deviation of the mean.   **7.** Lengths that are between 20.6 and 49.4 inches are within two standard deviations of the mean. The mean is 35, so the range is $35 - 2(7.2)$ or 20.6 to $35 + 2(7.2)$ or 49.4.   **9.** Sample answer: Both are calculated statistical values that show how each data value deviates from the mean of the data set. The mean absolute deviation is the mean of the absolute values of the differences between each number and the mean of the data set. The standard deviation shows how the data deviates from the mean of the data.

*Pages 715–716   Lesson 9-5   Extra Practice*

**11 a.** 11.41; Sample answer: The average distance each data value is from the mean is 11.41 miles per hour.   **11b.** Speeds that are between 89.23 and 117.03 miles per hour are within one standard deviation of the mean.   **13.** The mean absolute deviation is less than the standard deviation; The range is greater than the mean absolute deviation.   **15a.** South America   **15b.** Africa: mean: 1,517.7; median: 1,100; modes: 1,000 and 1,100; South America: mean: 1,461.05; median: 1,350; mode: 1,000   **15c.** Africa: the median or mode, since both of these measures are 1,100; South America: mean or median, since most of the data is near these measures   **15d.** Africa: range: 3,660; median: 1,100; $Q_3$: 2,145; $Q_1$: 850; interquartile range: 1,295; South America: range: 3,850; median: 1,350; $Q_3$: 1,869; $Q_1$: 978; interquartile range: 891

*Pages 721–722   Lesson 9-6   Independent Practice*

**1** The distribution is not symmetric. There is a cluster from 71–100 and a peak at the interval 81–90. The distribution has a gap from 61–70 percent. There are no outliers.   **3 a.** Sample answer: The distribution is not symmetric since the lengths of each box and each whisker are not the same. There is an outlier at 7.5.   **b.** Sample answer: The distribution is not symmetric. So, the median and interquartile range are appropriate measures to use. The data are centered around the median of $4. The spread of the data around the center is $1.25.   **5a.** The distribution in the top box plot is symmetric, so you would use the mean and the mean absolute deviation. The distribution in the bottom box plot is not symmetric, so you would use the median and the interquartile range.   **5b.** It is not possible to find mean and mean absolute deviation. It is possible to find the median and interquartile range.

*Pages 723–724   Lesson 9-6   Extra Practice*

**7** The shape of the distribution is symmetric. There are no clusters or gaps. The peak of the data is in the interval 31–40. There are no outliers.   **9a.** The shape of the distribution is not symmetric since the lengths of each box and each whisker are not the same. There are no outliers.   **9b.** The distribution is not symmetric. So, the median and interquartile range are appropriate measures to use. The data are centered around the median of 19 visitors. The spread of the data around the center is about 22.   **11a.** No   **11b.** Yes   **11c.** Yes   **13.** Sample answer: line graph   **15.** Sample answer: circle graph   **17.** Sample answer: histogram

*Page 727   Chapter Review   Vocabulary Check*

**1.** univariate data   **3.** bivariate data   **5.** five-number summary   **7.** line of best   **9.** symmetric   **11.** two-way table   **13.** mean absolute deviation

*Page 728   Chapter Review   Key Concept Check*

Sample answers are given.

Step 1 Draw the line.

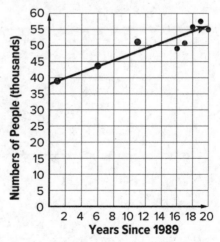

Step 2  Choose two points. (1, 39); (6, 44)
Step 3  Find the slope. $m = 1$
Step 4  Find the $y$-intercept. $b = 38$
Step 5  Write the equation of the line. $y = x + 38$

# Index

## Qq

## Rr

Name _____

Work Mats

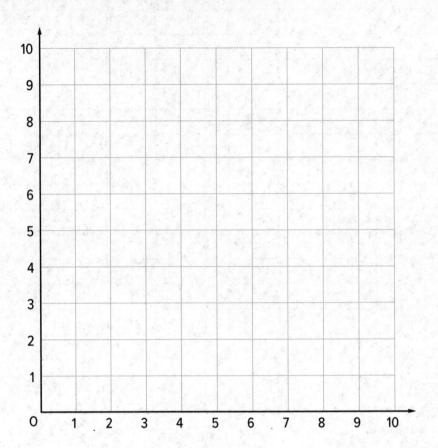

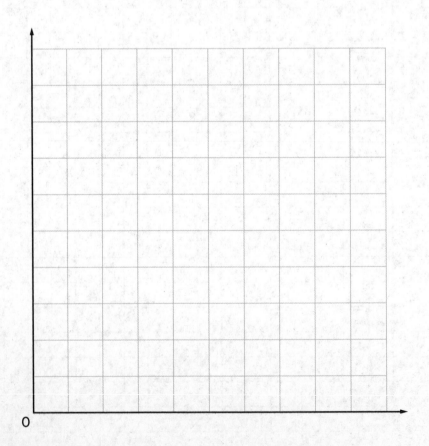

Name

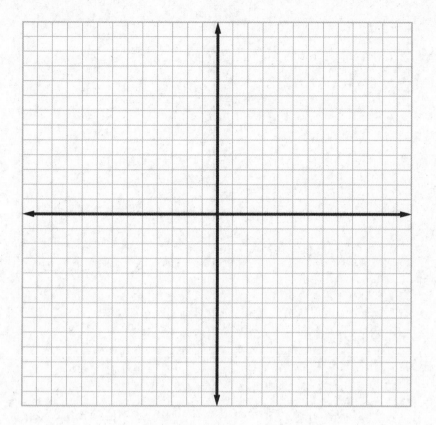

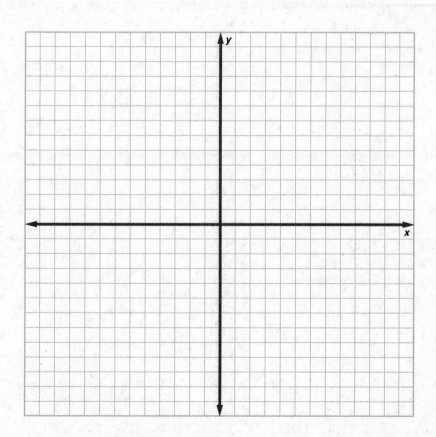

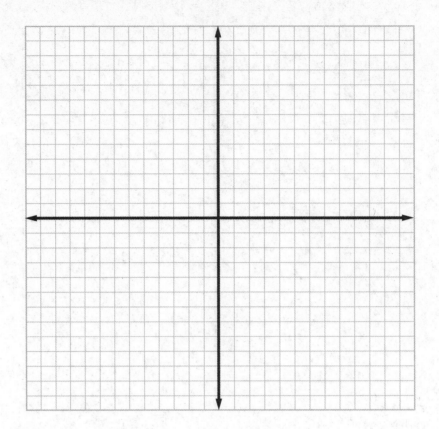

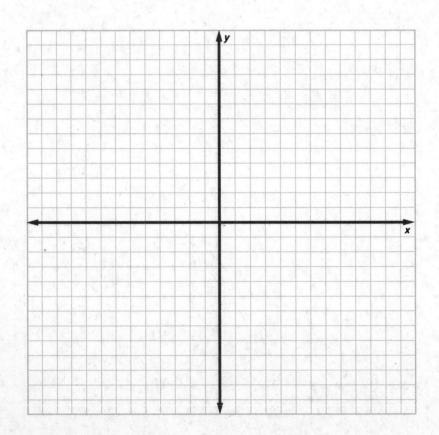

## What Are Foldables and How Do I Create Them?

Foldables are three-dimensional graphic organizers that help you create study guides for each chapter in your book.

**Step 1** Go to the back of your book to find the Foldable for the chapter you are currently studying. Follow the cutting and assembly instructions at the top of the page.

**Step 2** Go to the Key Concept Check at the end of the chapter you are currently studying. Match up the tabs and attach your Foldable to this page. Dotted tabs show where to place your Foldable. Striped tabs indicate where to tape the Foldable.

**Step 1**

**Step 2**

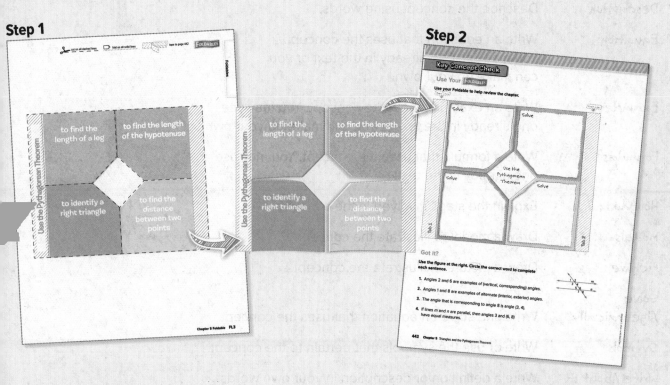

## How Will I Know When to Use My Foldable?

When it's time to work on your Foldable, you will see a Foldables logo at the bottom of the **Rate Yourself!** box on the Guided Practice pages. This lets you know that it is time to update it with concepts from that lesson. Once you've completed your Foldable, use it to study for the chapter test.

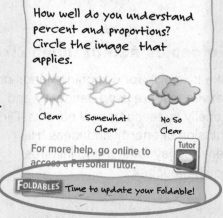

**Rate Yourself!**

How well do you understand percent and proportions? Circle the image that applies.

Clear      Somewhat      No So
           Clear          Clear

For more help, go online to access a Personal Tutor.

Tutor

FOLDABLES  Time to update your Foldable!

## How Do I Complete My Foldable?

No two Foldables in your book will look alike. However, some will ask you to fill in similar information. Below are some of the instructions you'll see as you complete your Foldable. **HAVE FUN** learning math using Foldables!

## Instructions and what they mean

| | |
|---|---|
| Best Used to... | Complete the sentence explaining when the concept should be used. |
| Definition | Write a definition in your own words. |
| Description | Describe the concept using words. |
| Equation | Write an equation that uses the concept. You may use one already in the text or you can make up your own. |
| Example | Write an example about the concept. You may use one already in the text or you can make up your own. |
| Formulas | Write a formula that uses the concept. You may use one already in the text. |
| How do I ...? | Explain the steps involved in the concept. |
| Models | Draw a model to illustrate the concept. |
| Picture | Draw a picture to illustrate the concept. |
| Solve Algebraically | Write and solve an equation that uses the concept. |
| Symbols | Write or use the symbols that pertain to the concept. |
| Write About It | Write a definition or description in your own words. |
| Words | Write the words that pertain to the concept. |

## Meet Foldables Author Dinah Zike

Dinah Zike is known for designing hands-on manipulatives that are used nationally and internationally by teachers and parents. Dinah is an explosion of energy and ideas. Her excitement and joy for learning inspires everyone she touches.

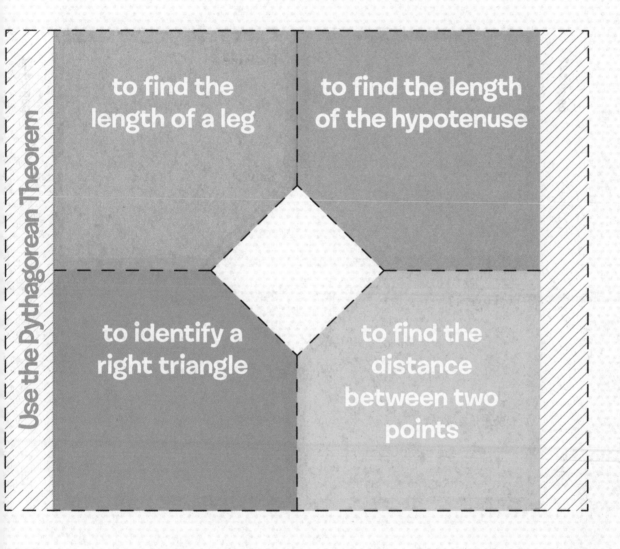

Use the Pythagorean Theorem

to find the length of a leg

to find the length of the hypotenuse

to identify a right triangle

to find the distance between two points

✂ cut on all dashed lines    ☐ fold on all solid lines    tape to page 442

**FOLDABLES**

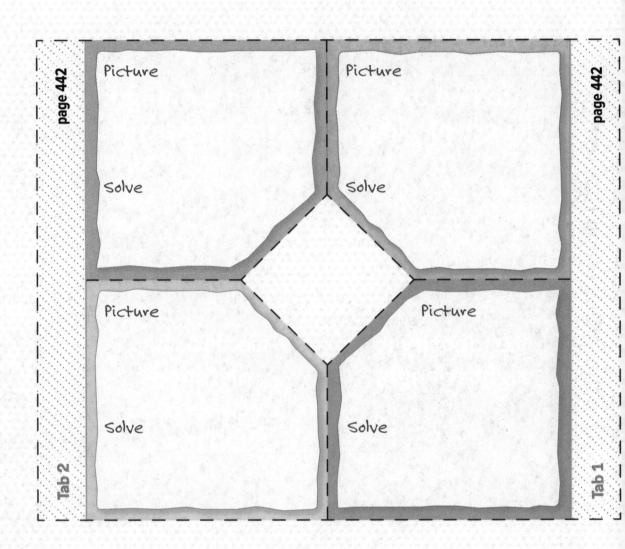

page 442

Picture

Solve

Picture

Solve

Picture

Solve

Picture

Solve

Tab 2

Tab 1

page 442

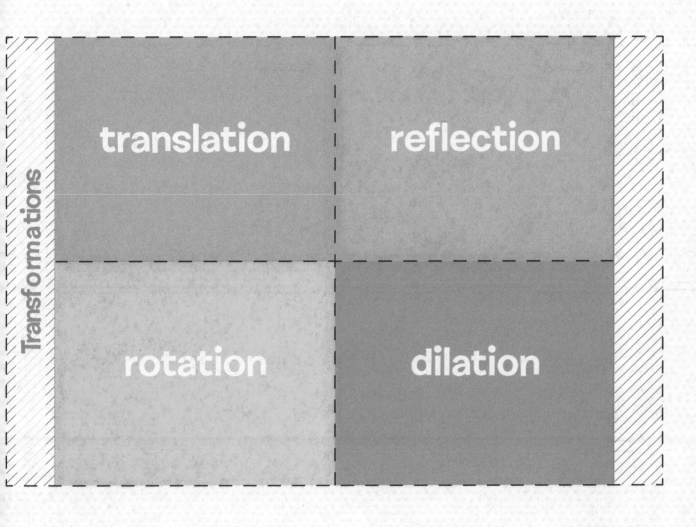

Transformations

translation

reflection

rotation

dilation

✂ --- cut on all dashed lines   ⬜▶ fold on all solid lines   ▨ tape to page 498   **FOLDABLES**

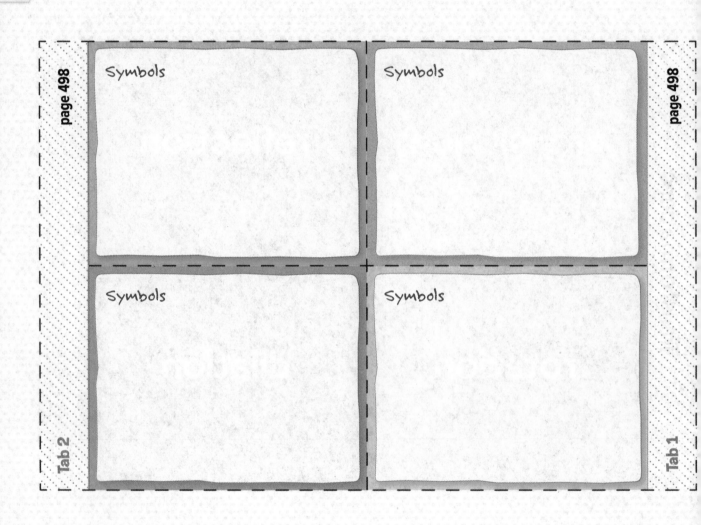

page 498

Symbols

Symbols

page 498

Symbols

Symbols

Tab 2

Tab 1

# Congruent Figures

| attributes | transformations |

| attributes | transformations |

# Similar Figures

✂ cut on all dashed lines     ▭ fold on all solid lines     tape to page 580     **FOLDABLES**

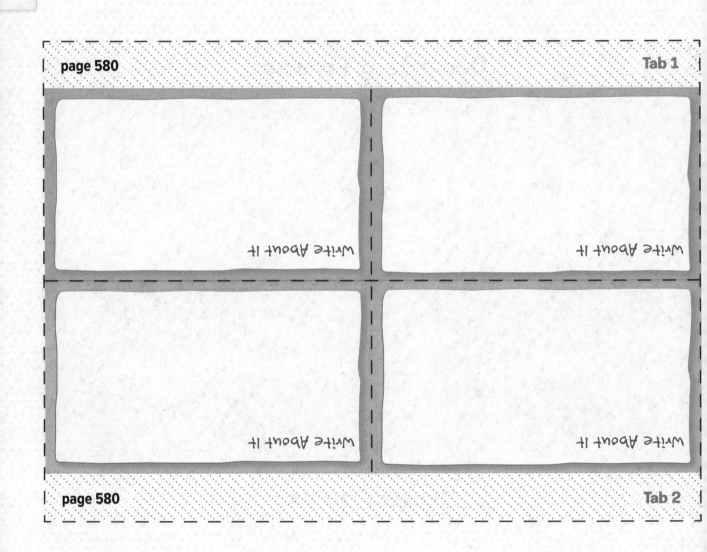

page 580                             Tab 1

Write About It

Write About It

Write About It

Write About It

page 580                             Tab 2

Volume

Surface Area

cylinder

cylinder

cone

cone

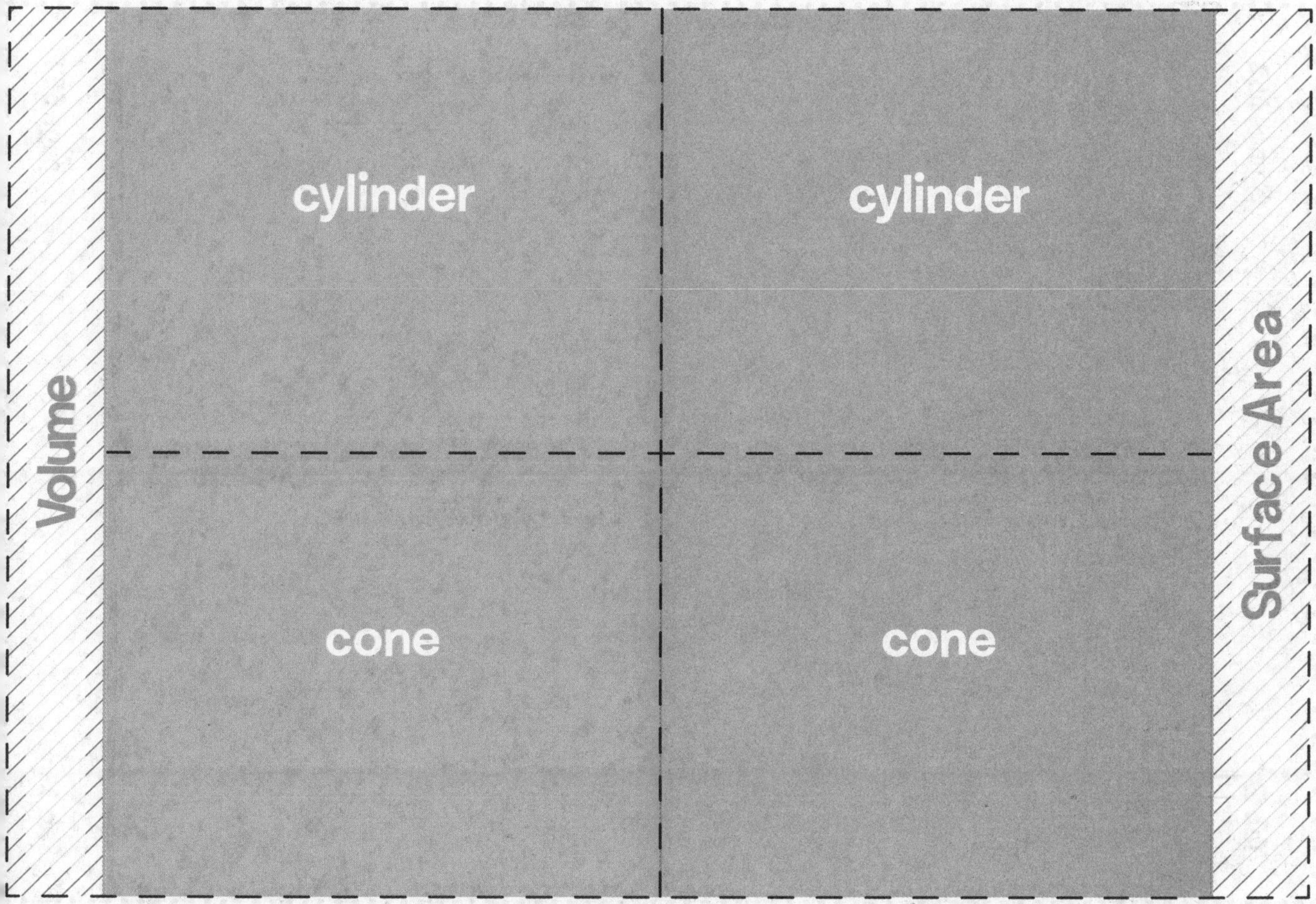

✂ cut on all dashed lines    fold on all solid lines    tape to page 652    **FOLDABLES**

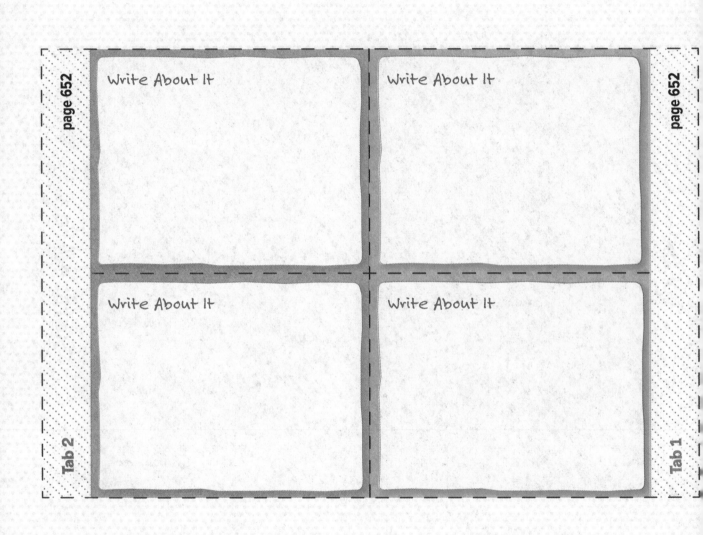

page 652

Write About It

Write About It

page 652

Tab 2

Write About It

Write About It

Tab 1

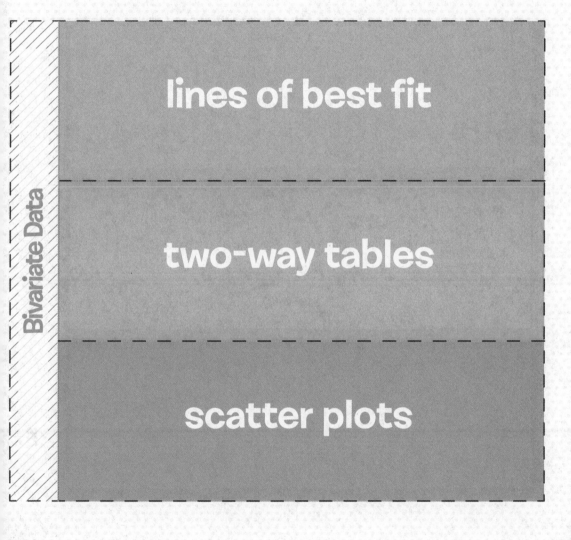

**Bivariate Data**

lines of best fit

two-way tables

scatter plots

cut on all dashed lines fold on all solid lines tape to page 728

FOLDABLES®

Foldables

Example

Example

Example

page 728